Lecture Notes in Biomathematics

Managing Editor: S. Levin

8

Charles DeLisi

Antigen Antibody Interactions

Springer-Verlag
Berlin · Heidelberg · New York

Lecture Notes in Biomathematics

Lecture Notes in Biomathematics

Managing Editor: S. Levin

8

Charles DeLisi

Antigen Antibody Interactions

Springer-Verlag
Berlin · Heidelberg · New York 1976

Author

Charles DeLisi
Department of Health, Education and Welfare
Public Health Service
National Institutes of Health
Bethesda, Maryland 20014/USA

Library of Congress Cataloging in Publication Data

DeLisi, Charles, 1941-
  Antigen antibody interactions.

  (Lecture notes in biomathematics ; vol. 8)
  Bibliography: p.
  Includes index.
  1. Antigen--antibody reactions. I. Title.
II. Series. [DNLM: 1. Antigen--Antibody reactions.
2. Models, biological. W1 LE334 v. 9 / QW570 D354a]
QR187.A1D44   612'.11822   76-15623

AMS Subject Classifications (1970): 92-00, 80A30, 82A30

ISBN 3-540-07697-2 Springer-Verlag Berlin · Heidelberg · New York
ISBN 0-387-07697-2 Springer-Verlag New York · Heidelberg · Berlin

CONTENTS

1

INTRODUCTION

1.1   Organization of the Immune System

One of the most important survival mechanisms of vertebrates is their ability to recognize and respond to the onslaught of pathogenic microbes to which they are continuously exposed.   The collection of host cells and molecules involved in this recognition-response function constitutes its immune system.   In man, it comprises about 10^{12} cells (lymphocytes) and 10^{20} molecules (immunoglobulins).   Its ontogenic development is constrained by the requirement that it be capable of responding to an almost limitless variety of molecular configurations on foreign substances, while simultaneously remaining inert to those on self components.   It has thus evolved to discriminate, with exquisite precision, between molecular patterns.

The foreign substances which induce a response, called antigens, are typically large molecules such as proteins and polysaccharides.   The portions of these with which immunoglobulins interact are called epitopes or determinants.   A typical protein epitope may consist of a configuration formed by the spatial arrangements of four or five amino acids and have an average linear dimension of about 20 Å.

The induction of a response is mediated by the interaction between epitopes and immunoglobulin molecules which function as receptors on lymphocyte membranes.   Prior to antigenic contact, lymphocytes are relatively small (5-8 microns in diameter) and probably in a resting phase of the cell cycle (G_0).   Polysomes are absent, mitochondria scarce, and there is no discernible endoplasmic reticulum.   Under appropriate circumstances contact leads to rapid proliferation and differentiation accompanied by changes in size (often a two-fold increase in diameter) and morphology (abundant cytoplasm, appearance of polysomes, conversion of heterochromatin to euchromatin, etc.).   This amplifed and highly differentiated population may effect elimination of antigen by both humoral (antibody-mediated) and cellular mechanisms.   The two possibilities reflect a functional bifurcation in the immune system which can be understood in terms of its ontogeny.

The precursors of immunologically active cells arise in the fetal yolk sac and lodge

2

in the adult bone marrow where they differentiate into various types of blood cells (Figure 1.1). Lymphocytes diversify further, becoming T cells if additional differentiation occurs in the thymus, and B cells if it occurs in the avian bursa or its mammalian equivalent. B lymphocytes are the precursors of antibody-secreting cells whereas T lymphocytes synthesize but do not secrete an appreciable quantity of antibody. The humoral response which is

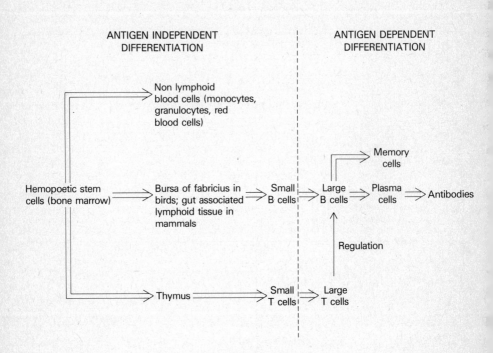

Figure 1.1

characterized by elevated concentrations of antibody diffusing in the blood serum therefore has its origin in the B cell population. Reactions to solutions of unaggregated large molecule are commonly of this type.

Many humoral responses can be intensified by repeated injection of the same antigen. The intensified (anamnestic) reaction is specific for the particular antigen used, i.e., primary immunization followed by challenge with a structurally unrelated antigen generally will not lead to a more pronounced reaction. It is of theoretical as well as practical importance however, that occasionally (with a probability of one in several hundred) an anamnestic response may be evoked by challenge with an antigen apparently unrelated to the primary

3

immunogen (Inman, 1974; Varga <u>et al</u>., 1974). In such instances the antigens are said to <u>cross react</u>.

Specificity is also a property of lymphocyte-mediated cellular immunity. Typical examples of this sort of response are the rejection of tissue grafts from genetically different donors and the destruction of tumor cells. In addition, T cells play an important role in the regulation of humoral responses to many antigens (see, for example, v.23 of <u>Transplant Rev</u>, 1975, G. Möller, ed.). Cellular responses of this type are to be distinguished from those mediated by phagocytic scavengers such as granulocytes and macrophages. The latter are not part of the lymphocytic system, lack its recognition capabilities, and respond entirely nonspecifically. For an introduction to cellular immunity the book by Eisen (1974) can be consulted. It will not be central to this volume, which will focus on a few conceptual aspects of the humoral response.

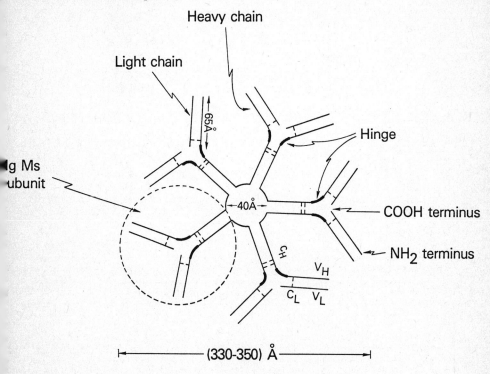

Figure 1.2. IgM has ten heavy and ten light polypeptide chains. Each heavy chain consists of a region (V_H) in which the amino acid sequence is peculiar to that particular molecule, and four other <u>constant</u> regions comprising C_H) which are homologous to one another and have a relatively invariant sequence. Each light chain contains one variable (V_L) and one constant (C_L) region.

1.2   Antibody Structure

Antibodies are globulin proteins (immunoglobulins) whose molecular compositic can be expressed as $(HL)_{2n}$ where H and L represent heavy and light polypeptide chain respectively, and n is an integer. Thus immunoglobulin G (IgG) for which n = 1 has tw heavy and two light chains; immunoglobulin M (n = 5) has ten (Figure 1.2). In both cas the light chain has a molecular weight of about 22,500 daltons and the heavy chain about 53,000.

Immunoglobulin V Regions. Aside from these physical characteristics, a variet of generalizations have emerged from amino acid sequence studies. For both IgG and Ig chains of the same weight have identical sequences in any single molecule. Different Ig (IgM) molecules, however, show marked differences in the sequence of the first 110 res dues, where counting begins at the amino (NH_2) terminus of the chain (Figure 1.2). T amino acids comprise the variable (or V) region.

Within this region there are a few subsequences which show extreme variabilit (Wu and Kabat, 1970). Since both local and long-range primary structure (sequence) expected to have an important influence on three-dimensional geometry, these hypervaria regions should confer distinctive and diverse structural features on antibodies; precise the properties needed for the immune system to be able to interact specifically with the la variety of antigens it may encounter. Wu and Kabat were thus led to speculate that the tial arrangement formed by the folding of the heavy and light chain hypervariable regio with one another is the antibody combining site, i.e., the site which binds epitope. Th prediction has now been verified by x-ray diffraction data at 2Å resolution, from sever different immunoglobulins (Segal et al., 1974; Edmundson et al., 1974).

All available evidence indicates that there is one combining site for each heavy light chain pair. Thus IgG with two heavy and two light chains has a valence of two; I a valence of ten. The phenomenological measure of the strength of the interaction betw combining site and epitope is the equilibrium constant or affinity of the reaction.

Immunoglobulin C Regions. The remaining amino acids in each chain show mu less variability and constitute the constant (or C) region. When the constant region of IgG heavy chain is compared with that from an IgM, striking dissimilarities are found b in amino acid sequence and chemical reactivity. Immunoglobulins which so differ are said to belong to different classes. There are in all five major immunoglobulin classes, but only the two which have been most extensively studied, IgG and IgM, will be considered in this volume.

One additional important aspect of sequence and structure is that a portion of the constant region of the heavy chain, the hinge, exhibits conformational flexibility, thus allowing relative lateral combining site motion. Important biological implications of this are beginning to emerge. The subject will be developed in Chapter 2 and will arise again in several subsequent chapters. For a good introduction to antibody structure, the book by Kabat (1968) is recommended.

1.3   The Theory of Clonal Selection

Attempts to understand the means by which antigen elicits a specific humoral response have led to a number of theories. The one currently accepted by immunologists emerged from the clonal selection hypothesis proposed by Burnet (1957) and Jerne (1955), which in outline says the following.

(1) Each B cell is phenotypically committed, prior to antigenic contact, to the production of antibodies which are homogeneous with respect to their affinity for antigen. The B cell population, however, is heterogeneous (Figure 1.3).

(2) A necessary event in the immune response is antigen-cell encounter which will, under appropriate circumstances, trigger differentiation, proliferation, and antibody production.

(3) The encounter is mediated by a set of mobile but membrane-bound antibody-like receptors which, on a given B cell, are homogeneous with respect to their affinity for antigen. Moreover, the affinity is identical to that of the antibodies ultimately secreted by the B cell progeny.

The heterogeneity in the B cell population postulated in (1) provides for the diversity which is necessary if the immune system is to respond to a wide variety of antigens. At the same time, by requiring phenotypic restriction of individual cells, presumably to the expression of just one light and one heavy chain variable region gene, a specific response is assured. The antigen-cell encounter in (2) provides a means of trapping antigen. Clearly

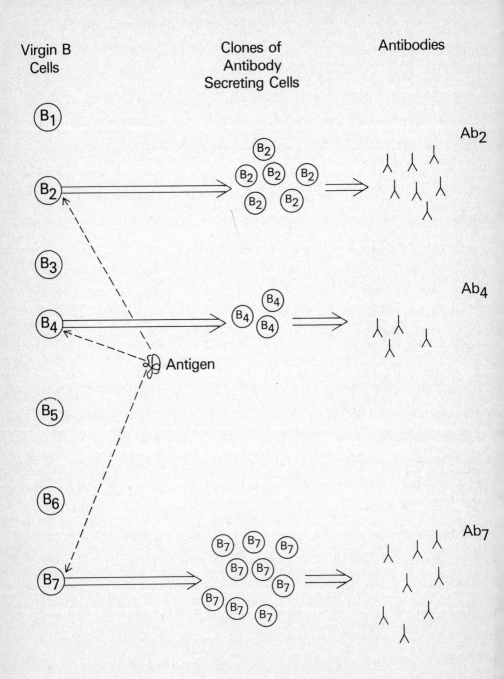

Figure 1.3. Antigen binds to a subpopulation of B lymphocytes (in this case B_1, B_4 and B_7) triggering clonal expansion and antibody production.

the energetics of the interaction plays an important role here in determining which cells are to be selected for stimulation. It also provides a means for amplification through clonal expansion. Moreover, the immunoglobulin products of the resulting clone must all interact favorably with the antigen because of (3). Since the theory predicts a spectrum of cells expressing different combining sites, a variety of distinct clones will be stimulated, giving rise to a population of antibodies with a spectrum of binding properties and the serum of the animal will therefore contain a heterogeneous antibody population (Figure 1.3).

The theory has had an enormous impact on immunological thinking and experimentation during the last fifteen years, and current research to a large extent involves providing it with a sound quantitative basis. Its importance insofar as the present volume is concerned resides in the role played by the energetics of the antigen-antibody interaction. A detailed consideration of the thermodynamics of the interaction, assuming only that antigen binding by receptor is requisite for stimulation, leads to a variety of semiquantitative predictions about the characteristics of the response (Siskind and Banaceraf, 1969). For example, if cellular selection is affinity-controlled, then as antigen is depleted during a response to the point where not enough remains to bind all cells, those whose receptors have a high affinity for antigen will be preferentially stimulated. Effects of this microevolutionary pressure for the preferential stimulation of high affinity cells as the response progresses have, in fact, been observed. However, although predictions of this sort are based only upon the thermodynamics of binding, no model of antigen cell encounter can be complete unless it distinguishes binding reactions from triggering reactions. This requirement is indicated by numerous experimental observations including the finding that binding of antigen to receptor may lead to paralysis (unresponsiveness to a subsequent, otherwise immunogenic dose of antigen) as well as stimulation.

A second requirement for stimulation must therefore be considered explicitly. One possibility with which many immunologists concur is that antigen receptor lattice formation may be necessary (Bell, 1975b). A theory for this will be presented in Chapter 8 and its immunological consequences pursued. It will be seen that the response can exhibit different behavioral modes depending on the values of the kinetic parameters. Thus, knowledge of the parameters is crucial for suggesting immunologically realistic behavior as well as for detailed verification of many of the predictions of clonal selection. For the same reason it is equally important to understand the biological factors which affect the parameters. In Chapters 3-6, the mathematical methods necessary to obtain this information from experin will be developed and applied, and some of the biological implications will be discussed Chapter 7. The first step in such an analysis or, indeed, in developing a theory of B ce triggering, is to understand the role played by geometry in influencing the physical behav and binding properties of antibodies. This will be the subject of Chapter 2.

2

HINGE AND VALENCE EFFECTS ON ANTIBODY BINDING PROPERTIES

2.1   General Considerations

Since the characterization of bond formation between large molecules must be phenomenological, a quantitative description of the reaction must be in terms of thermodynamic
parameters; in particular, free energies and equilibrium constants. In addition, however,
because an antigen is large compared to an antibody combining site and its structural features highly heterogeneous, a complete description of the interaction will require specifi-
cation of a set of free energies.

In general, an antigen is characterized by a set of epitopes $\{E_i\}$, i = 1, 2, ... k,
there being E_i of type i. Antigens for which E_i = 1 for all i are said to be <u>univalent</u>. An
antigen is multivalent if E_i > 1 for at least one value of i. Because of epitope heterogeneity,
<u>in vitro</u> studies generally do not use natural antigen. Instead, an antigen is chemically mod-
ified by covalently coupling some ligand, H, to its surface to produce a molecule, H_nX,
where X represents the antigen and n is the number of ligands per antigen molecule. Typically the ligands have molecular weights of several hundred daltons. If immunization is with
H_nY, then antibodies will be produced against H and the determinants of Y. When X and Y
are non cross reacting, then H_nX will be homogeneous with respect to the antiserum, <u>i.e.</u>,
in a mixture of antiserum and H_nX, only H will bind antibodies. The <u>carrier molecule</u>, Y
(which may be any antigen other than X), must be used in the immunization because small
ligands alone do not elicit an immune response. Small molecules which bind antibody but
do not by themselves stimulate a response are called <u>haptens</u>. In general, I will continue to
use <u>epitope</u> to describe haptens that are covalently bound to antigen, and reserve <u>ligand</u>
for those that are not.

Let G be defined as the free energy change accompanying bimolecular reaction between antibody combining site and ligand, and let K_1 be the corresponding equilibrium constant. These may not be quite the same as the free energy and equilibrium constant for
binding the identical epitope since the collision frequency between the reactive units may
be higher in the former case than in the latter. However, if the reaction rates are not dif-

fusion limited, the difference should be small (see Chapter 8) and for the present it will be ignored.

It is evident that G will make an important contribution to the strength of the inter action between an antibody molecule and an antigen. However, aside from this, the geo-metrical and structural features of the antigen and antibody play an important role.

The two relevant characteristics of an antibody in this regard are hinge flexibilit (Figure 1.2) and multiple combining sites. These properties allow multiple bonding betwe antigen and antibody. This is generally referred to as multivalent attachment, or in the cas of IgG to which the analysis will be limited, bivalent attachment. Bivalent binding can occu of course, only with antigens having a sufficiently dense array of repeating determinants but since these are common, such binding should be regarded as typical and it would be surprising if it were not biologically significant.

When one considers the physical chemistry of the interaction, a potential advantage of bivalent binding is immediately apparent. Since the ratio of bound to unbound molecule is proportional to the equilibrium constant, and that depends exponentially on the free en ergy, a small change in the latter as a consequence of forming a second bond, may be am-plified into a substantial shift in the concentration of bound molecules. More specifically, the equilibrium constant for a combining site-epitope interaction is

$$K_1 = \exp{(G/RT)} \qquad\qquad 2.1$$

where RT is the thermal energy, R being the gas constant, and T the temperature. With G expressed in calories per mole, R is in calories/(mole-$^\circ$K) and T is in $^\circ$K. I have adopte the convention that the interaction becomes more favorable as G increases. Typically, at room temperature, G/RT ranges from about 10 to 20, so substantial amplification is, in fact possible under ordinary conditions.

2.2   The Two-Particle Probability Density Function

The actual shift in the concentration of bound antibodies will depend upon the mag-nitude of the gain in free energy accompanying the formation of the second bond. In order to estimate this, consider the reaction between bivalent antibody and bivalent antigen as shown in Figure 2.1.

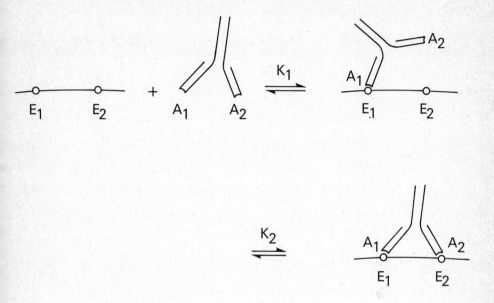

Figure 2.1

The antigen may be thought of as a relatively large object such as a cell. Epitopes E_1 and E_2 are considered chemically identical, as are the antibody combining sites, A_1 and A_2. In the first reaction step, single site attachment between A_1 and E_1 is established. In the second step, A_2 binds intramolecularly to E_2. Let K_2 be the equilibrium constant for this step. The problem is to calculate K_1/K_2 as a function of molecular parameters.

It is apparent that in general the magnitude of an equilibrium constant will depend upon the reactivity of the chemical groups and the probability that they will assume positions which allow complexation. The former is, by definition, the same for the bimolecular and intramolecular reaction, whereas the latter is not. K_1/K_2 is therefore assumed to be independent of chemical reactivity but highly dependent upon any constraints imposed upon the motion of A_2 relative to A_1.

In order to make this idea precise, consider the general geometrical requirements for bond formation between a combining site A and an epitope E. Let coordinate systems be drawn through the center of mass of each as shown in Figure 2.2. Suppose that bond

formation requires that A be located within a small volume δv at the end of $\vec{R}^*$ in the coor-

dinate system of E. A further requirement is that the coordinates of system A be at angles

θ_1^*, θ_2^*, θ_3^* relative to the coordinate system of E. The permissible variation in the angu-

lar orientation of A is the solid angle $\delta \omega$ and the rotational interval $\delta \alpha$ about the z' axis.

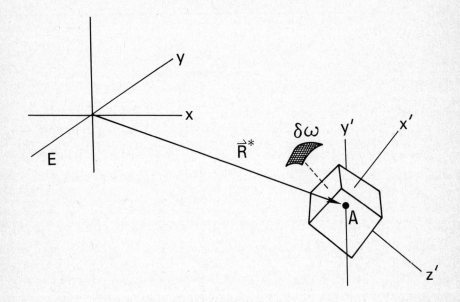

Figure 2.2

The spatial part of the two-particle distribution function, $S(\vec{R})\delta v$, is defined as the

probability of finding the center of mass of A within δv at the end of $\vec{R}$. $S(\vec{R})$ is, therefore,

the probability per unit volume that A's center of mass is at a particular point in the coordin-

ate system of E. In addition, one can define the angular part of the distribution function by

letting $T(\theta_1, \theta_2, \theta_3)\,\delta\omega\delta\alpha$ be the probability that A is within $\delta\omega$ and $\delta\alpha$ of the orientation

specified by θ_1, θ_2, θ_3.

If K_0 is the equilibrium constant for bond formation between particles containing

A and E when the two are within the required limits δv and $\delta\omega\delta\alpha$ of the location specified by $\vec{R}^*, \theta_1^*, \theta_2^*, \theta_3^*$, then,

$$K = S(\vec{R}^*) \, \delta v \, T(\theta_1^*, \theta_2^*, \theta_3^*) \, \delta \omega \, \delta \alpha \, K_0 \qquad\qquad 2.2$$

where the distribution functions S and T must be evaluated in some standard reference state, i.e., at a particular concentration. Since equation 2.2 is independent of the type of reaction, it follows that (DeLisi and Crothers, 1972)

$$\frac{K_1}{K_2} = \frac{S_1^* \, T_1^*}{S_2^* \, T_2^*} \qquad\qquad 2.3$$

where arguments have been omitted for notational simplicity.

These ideas can also be applied to the discussion of rate constants. Let k_1 and k_{-1} be the forward and reverse rate constants for the bimolecular reaction and let k_2 and k_{-2} be similarly defined for the intramolecular process, i.e.,

$$K_1 = k_1/k_{-1} \qquad\qquad 2.4$$

and

$$K_2 = k_2/k_{-2} \qquad\qquad 2.5$$

For the units to react, a translational activation entropy barrier must be overcome. This means, essentially, that they must "find" each other and attain positions appropriate for reaction. The chance of this happening can, as was shown, be evaluated from knowledge of the distribution functions. Once correct positioning is achieved, complexation is assumed to occur with an intrinsic association constant, k_0, which is independent of the order of the reaction (i.e., bimolecular or intramolecular). Translational activation entropy for dissociation, on the other hand, should be close to zero (DeLisi and Crothers, 1973) so the distribution functions play no role in the reverse reaction. Thus

$$k_1 = S_1^* \, \delta v \, T_1^* \, \delta \omega \, \delta \alpha \, k_0 \qquad\qquad 2.6$$

$$k_2 = S_2^* \, \delta v \, T_1^* \, \delta \omega \, \delta \alpha \, k_0 \qquad\qquad 2.7$$

In the absence of strain, $k_{-2} = k_{-1}$ and the equilibrium expression, equation 2.3, can be recovered. Finding the kinetic and thermodynamic parameters is therefore reduced to obtaining appropriate distribution functions.

In the case of a simple bimolecular reaction in which A_1 and E_1 interact by speci fic bond formation, but are otherwise completely independent, S_1 is constant over the en tire solution volume. Therefore,

$$S_1 = N/V \qquad\qquad 2.8$$

where N is the number of A particles per volume V in the standard state (1 mole/liter). Moreover, the function T_1 is of interest when the centers of mass of the reactive units are separated by $\vec{R}^*$. In such a case, the normalization condition is

$$\tfrac{1}{2} \int_0^{4\pi} d\omega \int_0^{2\pi} T_1 \, d\alpha = 1 \qquad\qquad 2.9$$

where the factor $\tfrac{1}{2}$ arises because only the angular positions for which the sites are in the upper half of the plane (i.e., outside the antigenic surface) are permitted. Since T_1 is constant in this case

$$T_1 = \frac{1}{4\pi^2} \qquad\qquad 2.10$$

Therefore, for the bimolecular step, equation 2.2 becomes

$$K_1 = \frac{N\,\delta\nu\,\delta\omega\,\delta\alpha\,K_0}{4\pi^2 V} \qquad\qquad 2.11$$

and from equation 2.3

$$\frac{K_1}{K_2} = \frac{N}{4\pi^2\,V\,S_2^*\,T_2^*} \qquad\qquad 2.12$$

The problem is to evaluate the spatial and angular parts of the probability densi function governing the distance between A_2 and E_2. The behavior of this function is in timately connected with the configurational details of the polypeptide chain intervening t tween the antibody combining sites. This is illustrated schematically in Figure 2.3. Th angles θ_i, $i = 1, 2 \ldots$ are fixed and so are the bond lengths, but each torsional angle ϕ_i ca

assume any of a discrete range of values with weights W_{ij} where j labels rotational positions.
It is clear that for a certain set,$\{\phi_i\}$, the chain may be extended as shown in Figure 2.3,
whereas for another set it may be relatively compact. The probability that the two ends are
at a particular distance will therefore depend upon the range of values allowed to each ϕ_i and
the weights W_{ij}, and these are both intrinsic properties of the amino acid sequence.

The problem is a generalized random walk in which the allowed steps are subject
to arbitrary constraints. If the number of steps (bonds) become large enough and the conditions of the central limit theorem are satisfied, the spatial part of the distribution function tends to become gaussian. However, for a finite number of steps, an exact solution has
only been found for one case, _viz._, for completely free rotation about each joint (Raleigh,

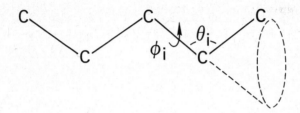

Figure 2.3. A schematic of a polypeptide backbone segment. C represents a carbon atom
and the straight lines are bonds. θ_i (i = 1, 2 ...) is the angle between bond i and i + 1
and has a fixed constant value for each pair. However, the torsional angle ϕ_i, which labels
rotation about bond i, may assume any of a discrete set of values, each having some weight
attached to it. Different sets $\{\phi_i\}$ will therefore generally lead to different distances between two particular atoms.

1919). Finding exact distribution functions when restrictions are present is one of the
most important classes of unsolved problems in the statistical mechanics of chain molecules.

Although it is possible to find moments exactly (Flory, 1968) and reconstruct the
distribution approximately, using, for example, Hermite polynomials as a basis set (Grad,
1949; Jernigan, 1967), this approach will not be followed since the required structural data
are lacking. Instead, a qualitative analysis will be presented, based on a simple though
successful model developed by Crothers and Metzger (1972).

2.3 The Crothers-Metzger Model

The essence of the model is to take full cognizance of our ignorance by replacing all unknown structural features by universal joints. Thus the hinge is as shown in Figu 2.4. In addition, a rigid bond links epitope to combining site, but the epitope is suppose to be attached to the surface by a completely flexible joint. This last supposition, of cours is not meant to represent reality since attachment is actually rigid. It is merely an arti- fice for obtaining the spatial distribution function and will not have an important effect on final result if all positions are considered to lead to intramolecular bond formation with equ likelihood.

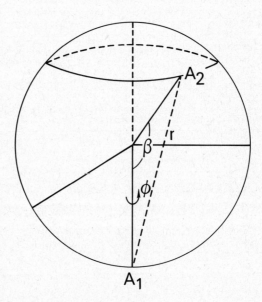

Figure 2.4. Antibody combining sites A_1 and A_2 subtend an angle β at the hinge.

The introduction of universal joints implies the assignment of equal weights to a accessible positions of the antibody combining sites. With free rotations allowed, the ori tations of the sites are again uncorrelated and the angular part of the distribution functio is given by

$$T_2^* = \frac{1}{4\pi^2}$$

2.13

With regard to S_2, Crothers and Metzger argue that within the framework of the model, little numerical error is introduced by taking it constant within a sphere of radius $< r >$, the mean distance between combining sites, and zero elsewhere. Then, since the probability density is normalized to one

$$S_2 = S^* = \frac{3}{4\pi <r>^3} \qquad r \leq <r> \qquad\qquad 2.14$$

Combining equations 2.12 - 2.14, one has that,

$$\frac{K_1}{K_2} = \frac{4\pi <r>^3 N}{3V} \qquad\qquad 2.15$$

An explicit expression for $<r>$ can be obtained by evaluating

$$<r> = \int_0^{\beta max} r P(\beta) \, d\beta \qquad\qquad 2.16$$

where $P(\beta)$ is the probability that two combining sites subtend an angle β at the hinge, i.e.,

$$P(\beta) = \frac{\sin \beta}{\int_0^{\beta max} \sin \beta d\beta} = \frac{\sin \beta}{1 - \cos \beta max} \qquad\qquad 2.17$$

In addition (Figure 2.4)

$$r = 2b \sin (\beta/2) \qquad\qquad 2.18$$

where b is the long axis of the Fab fragment (the distance from combining site to hinge).

Substituting equations 2.17 and 2.18 into equation 2.16 and integrating, one finds that

$$<r> = \frac{2b}{1 - \cos \beta max} \left\{ \sin (\beta max/2) - \frac{1}{3} \sin (3 \beta max/2) \right\} \qquad\qquad 2.19$$

For a completely flexible hinge, $\beta max = \pi$ and

$$<r> = 4b/3 \qquad\qquad 2.19a$$

Complete flexibility is, of course, unlikely. An experimental lower bound on βmax based on fluorescence experiments is 33° (Yguarabide et al., 1970). With this value

$$<r> = 0.174\ b \qquad\qquad\qquad 2.19b$$

These results apply to bivalent antibody interacting with bivalent antigen whose epitopes are separated by a distance less than $<r>$. If the antigen has many epitopes, results may be generalized by introducing an average density, x. Then if a is the area the antigen, the bimolecular and intramolecular equilibrium constants are given by

$$K'_1 = a\ x\ K_1 \qquad\qquad\qquad 2.20$$

and

$$K'_2 = \pi <r>^2\ x\ K_2 \qquad\qquad\qquad 2.21$$

Therefore

$$\frac{K'_1}{K'_2} = \frac{4\,a<r>N}{3V} \qquad\qquad\qquad 2.22$$

In terms of rate constants

$$K'_1 = \frac{k_1\ x\ a}{k_{-1}} \equiv \frac{k'_1}{k_{-1}} \qquad\qquad\qquad 2.23$$

and

$$K'_2 = \frac{k_2\ \pi <r>^2\ x}{k_{-2}} \equiv \frac{k'_2}{k_{-2}} \qquad\qquad\qquad 2.24$$

2.4 Analysis of Experiments

When one attempts to develop mathematical models, whether it be with regard t theories of the immune response or analysis of in vitro assays, it is important to have so idea of the relative magnitudes of kinetic parameters. Since they determine the time sca for different types of reactions, strong biases may be produced in favor of or against ce tain pathways, depending upon their values. This means that, according to equations 2 and 2.24, changes in such things as amino acid sequence (which affects k_2), fluidity of the medium in which reaction is occurring (which affects k_1 and k_2), and epitope densit (x), may lead to important qualitative changes in the behavior of an antigen-antibody system. Some of these ideas will be illustrated in subsequent chapters. The objective here is to obtain estimates of the kinetic and thermodynamic parameters by analyzing the appropriate experiments.

In a classic study, Hornick and Karush (1969, 1972) examined the effect of valence on binding reactions by comparing the phage neutralizing capabilities of antibodies of different valence. Briefly, the experimental protocol involves mixing antihapten antibodies [the hapten was dinitrophenyl (DNP)] with phage having DNP covalently coupled to its coat protein. The binding reaction neutralizes the phage so that it is no longer able to lyse bacteria. Thus, if aliquots of the mixture are taken at various times and tested for bacteriolysis by plaque formation, the results (number of plaques as a function of time) should reflect kinetic aspects of the antibody phage interaction. It is not known a priori, however, whether there is one sensitive site on the phage surface which, when bound, leads to loss of bacteriolytic capacity, or whether there are several sites.

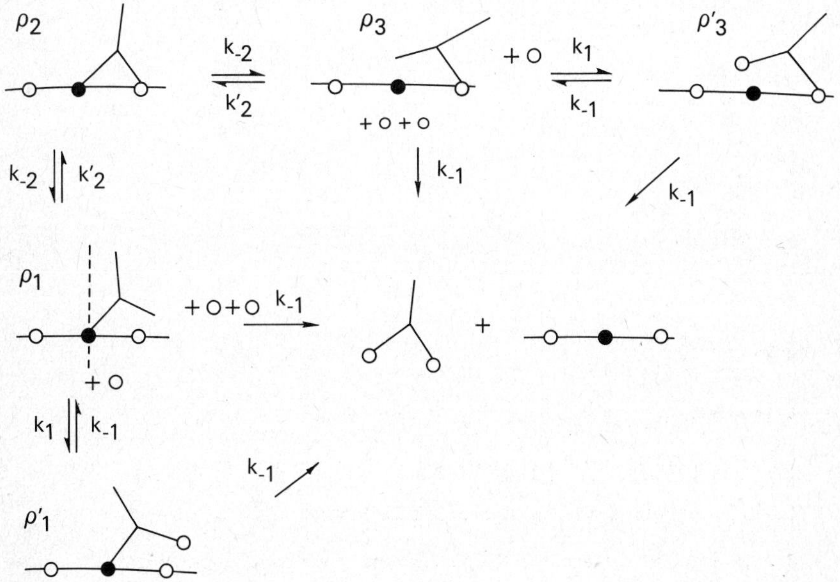

Figure 2.5. ρ_i are the concentrations of complexes shown. For example ρ_2 is the concentration of antibodies having one site bound to the sensitive phage site (●) and the other bound to an adjacent site (O). The phage bound hapten and free hapten are identical.

Neutralization is reversible, i.e., the phage regains its lytic capacity when the an[t]body dissociates. The reactivation can provide information about antibody dissociation ra[te] as well as the number of sensitive sites per phage. Since this information is required for analysis of neutralization, reactivation will be considered first.

In order to drive dissociation of the antibody, the solution of neutralized phage i[s] mixed with a concentration of free DNP groups several orders of magnitude higher than th[at] of epitope. Thus, any antibody that dissociated from the phage will bind free ligand rath[er] than reassociate (Figure 2.5). When only a single site becomes free, i.e., when ρ_1 is form[ed] the preferred subsequent event is not obvious. The antibody may, of course, dissociate fr[om] the phage. Alternatively its free site may bind free ligand, or it may again react intramo[le]cularly.

For the last possibility, there is a question as to the group with which the free s[ite] reassociates. I will assume that intramolecular reaction can only be with nearest neighb[or] so that the combining site reacts with the group from which it dissociated. The nearest ne[igh]bor assumption is based upon the estimated relationship between the mean combining site spacing and the mean epitope spacing. With 100 epitopes distributed over an area of 2×10^{-11} cm^2 (Crothers and Metzger, 1971), the average spacing is 5×10^{-7} cm. On the other hand, $<r>$ calculated from equations 2.19a and 2.19b is 8.7×10^{-7} cm and 1.1×10^{-7} cm, respectively, so a nearest neighbor approximation seems reasonable.

The rate constant for intramolecular reassociation is taken to be k_2'. This, it will be recalled, describes intramolecular reaction when there is a continuous determinant den[si]sity, so that regardless of which direction the bond takes in the bimolecular step, the chance of forming a second bond is the same. In other words, it is applicable when intra[a]molecular reaction is independent of rotation of the plane formed by the arms of the antibody about the dashed line in Figure 2.5. This is the same as saying that the directional re[]quirement on this plane has been met, and that is the basis for identifying the rate constant for the $\rho_1 \rightarrow \rho_2$ transition with k_2'.

The above considerations also apply to ρ_3 and ρ_3'. It is therefore apparent that $\rho_3 = \rho_1$ and $\rho_3' = \rho_1'$. Consequently, the equations of the model are

$$\frac{d\rho_1}{dt} = - (k_{-1} + k_1 H + k_2') \rho_1 + k_{-1} \rho_1' + k_{-2} \rho_2 \qquad 2.25$$

$$\frac{d\rho_1'}{dt} = k_1 H \rho_1 - 2k_{-1} \rho_1' \qquad 2.26$$

$$\frac{d\rho_2}{dt} = k_2' \rho_1 - 2k_{-2} \rho_2 + k_2' \rho_3 \qquad 2.27$$

As seen in Figure 2.5, ρ_1 is the concentration of antibodies having one site bound to the sensitive phage epitope, and the other free. The first term on the right in equation 2.25 describes its disappearance as the result of dissociation from phage, association between its free site and inhibitor, and intramolecular reaction, respectively. The second term represents its formation as the result of ligand dissociating from ρ_1', and the third, its formation as the result of intramolecular dissociation of ρ_2. Equations 2.26 and 2.27 are written by similar considerations. These are the only equations necessary since ρ_3 is symmetric to ρ_1, and ρ_3' is symmetric to ρ_1'.

Some insight into the kinetics can be gained by considering the magnitudes of the parameters. Since $k_{-1} \stackrel{\sim}{=} (1\text{--}100) \sec^{-1}$ (Froese et al., 1962; Day et al., 1963; Pecht, et al., 1972) and $k_1 H$ must also be rapid, it is reasonable to suppose that ρ_1' relaxes to equilibrium rapidly on the time scale of a 35-minute experiment. Therefore

$$\rho_1' \stackrel{\sim}{=} \frac{K_1 H}{2} \rho_1 \qquad 2.28$$

Consequently the kinetic model can be approximated by a two-variable system; viz.,

$$\frac{d\rho_1}{dt} = - (k_{-1} + \frac{k_1 H}{2} + k_2') \rho_1 + k_{-2} \rho_2 \qquad \rho_1 (0) = 0 \qquad 2.29$$

$$\frac{d\rho_2}{dt} = 2k_2' \rho_1 - 2k_{-2} \rho_2 \qquad \rho_2 (0) = 0 \qquad 2.30$$

Defining

$$\underset{\sim}{M} \equiv \begin{pmatrix} m_{11} & m_{12} \\ m_{21} & m_{22} \end{pmatrix} = \begin{pmatrix} - (k_{-1} + k_1 H/2 + k_2') & k_{-2} \\ 2k_2' & -2k_{-2} \end{pmatrix} \qquad 2.31$$

the solutions to equations 2.29 and 2.30 can be written as

$$\rho_1 = \frac{\rho_0 \, m_{12}}{\mu_r^{(+)} - \mu_r^{(-)}} \left[\exp(\mu_r^{(+)} t) - \exp(\mu_r^{(-)} t) \right] \qquad 2.32$$

and

$$\rho_2 = \frac{\rho_0}{\mu_r^{(+)} - \mu_r^{(-)}} \left[(\mu_r^{(+)} - m_{11}) \exp(\mu_r^{(+)} t) - (\mu_r^{(-)} - m_{11}) \right. \qquad 2.33$$

$$\left. \exp(\mu_r^{(-)} t) \right]$$

where ρ_0 is the total concentration of sensitive sites, and

$$\mu_r^{(\pm)} = \left(\frac{m_{11} + m_{22}}{2} \right) \left[1 \pm \sqrt{1 - \frac{4|M|}{(m_{11} + m_{22})^2}} \right] \qquad 2.34$$

are the eigenvalues of M, and $|M|$ its determinant.

It is known from equilibrium studies that the concentration of bivalent antibodie needed to neutralize phage is several orders of magnitude smaller than the amount of mor valent antibody used to obtain the same extent of neutralization. This implies that

$$K_2' \equiv \frac{k_2'}{k_{-2}} \gg 1$$

Consequently the expansion of equation 2.34 can be well approximated by keeping only t up to first order in $\dfrac{|M|}{(m_{11} + m_{22})^2}$. Thus

$$\mu_r^{(+)} = m_{11} + m_{22} - \mu_r^{(-)} \qquad 2.35$$

$$\mu_r^{(-)} = \frac{|M|}{m_{11} + m_{22}} \qquad 2.36$$

Moreover, the same considerations imply that $|\mu_r^{(+)}| \gg |\mu_r^{(-)}|$, so the expon tial in $\mu_r^{(+)}$ is a rapid transient, and the observed kinetics are controlled by $\mu_r^{(-)}$. T fraction of sensitive sites F that are bound at time t is then

$$F = \frac{P_1 + P_2}{P_0} \cong \exp(\mu_r^{(-)} t) \qquad \qquad 2.37$$

Consequently if there are n sensitive sites per phage, the fraction P reactivated at time t is

$$P = (1 - F)^n \qquad \qquad 2.38$$

or

$$\ln(1 - P^{1/n}) = \mu_r^{(-)} t \qquad \qquad 2.39$$

The best straight line fit to this curve is obtained for $n = 1$, and at this value

$$\mu_r^{(-)} = -3.3 \times 10^{-4} \text{ sec}^{-1}$$

It therefore seems likely that, as indicated by Hornick and Karush, there is only one sensitive site per phage. Without any additional information, however, the magnitudes of the rate constants for the intramolecular process cannot be inferred.

It is useful to consider situations in which relaxation reflects the rate constant of just a single elementary step. There are two limits of interest. First with $k_1 H \gg k_2'$

$$\mu_r^{(-)} \equiv -k_r = -2k_{-2} \qquad \qquad 2.40$$

Alternatively, if $k_2' \gg 2k_{-1} + k_1 H$

$$k_r = \frac{2k_{-1}}{K_2'} \qquad \qquad 2.41$$

where the notation k_r has been introduced for the effective reverse rate constant.

According to equation 2.40, reactivation is rate limited by breakage of the intramolecular bond. Once this happens, the antibody dissociates immediately and both sites are blocked by ligand, preventing subsequent reassociation. According to equation 2.41 on the other hand, intramolecular dissociation is most often followed by reassociation. Occasionally, however, the second bond breaks before intramolecular reaction can recur and when this happens, inhibitor blocks both sites. A discussion of which equation might be more appropriate to the Hornick-Karush experiments is deferred until after development of the neutralization analysis.

Neutralization may occur on the first step of the antibody phage interaction by monovalent contact with the sensitive site, and the resulting complex (ρ_1) may then be s[...] bilized by formation of an intramolecular bond. Alternatively, the antibody may bind ad[...] jacent to the sensitive site and then neutralize the phage when the second bond is estab- lished (Figure 2.6). The rate constant for forming the second bond, k_2'', will be less th[...] k_2' since an adjacent site will not always be free. However, if neutralization has not pro[...] gressed too far, the difference should not be large (e.g., with about 50% of phage epitop[...] bound k_2''/k_2' is not expected to differ significantly from $\frac{1}{2}$). Thus the equations corre- sponding to the kinetic model in Figure 2.6 are

$$\frac{d\rho_1}{dt} = 2k_1 c\,(\rho_0 - \rho_1 - \rho_2) - \rho_1\,(k_{-1} + k_2'') + k_{-2}\,\rho_2 \qquad 2.42$$

$$\frac{d\rho_2}{dt} = k_2''\,\rho_1 - 2k_{-2}\,\rho_2 + k_2''\,\rho_3 \qquad 2.43$$

$$\frac{d\rho_3}{dt} = 2k_1 c\,(\rho_0 - \rho_3 - \rho_2) + k_{-2}\,\rho_2 - (k_{-1} + k_2'')\,\rho_3 \qquad 2.44$$

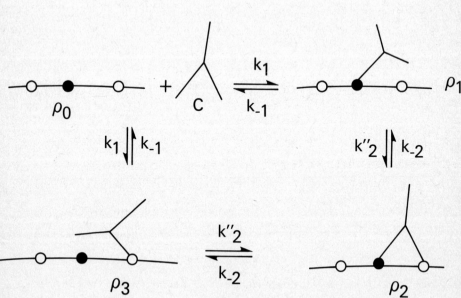

Figure 2.6

Equation 2.44 merits some comment. ρ_3 is the concentration of antibodies singly bound adjacent to the sensitive site <u>and</u> appropriately oriented to neutralize the phage by intramolecular bond formation. Thus if there are m adjacent sites, all terms in the first parenthesis on the right should be multiplied by m. However, the forward rate constant for the bimolecular step will be reduced by a factor of m since the formation of ρ_3 requires appropriate orientation. Defining

$$\underset{\sim}{M} = \begin{pmatrix} -(2k_1 c + k_2'' + k_{-1}) & 2(k_{-2} - 2k_1 c) \\ \\ k_2'' & -2k_{-2} \end{pmatrix} \qquad 2.45$$

and $\quad \xi = \rho_1 + \rho_3 \qquad\qquad\qquad 2.46$

the equations of the model can be written

$$\frac{d}{dt}\begin{pmatrix} \xi \\ \rho_2 \end{pmatrix} = \underset{\sim}{M}\begin{pmatrix} \xi \\ \rho_2 \end{pmatrix} + \begin{pmatrix} 4 k_1 c \rho_0 \\ 0 \end{pmatrix} \qquad 2.47$$

Since the neutralization process has a half-life of about 14 minutes, it can be argued by again taking account of the magnitudes of the known parameters that the observed relaxation is controlled by

$$\mu_f^{(-)} = -\frac{2(2k_1 k_2'' c + k_{-2} k_{-1})}{(k_2'' + k_{-1})} \qquad 2.48$$

where $\mu_f^{(-)}$ is the smallest eigenvalue of the matrix defined by equation 2.45. Thus, in general, the effective forward rate constant is a function of the rate constants for all the elementary steps. However, it will be useful to consider limiting expressions. In the limit, as $k_{-1}/k_2'' \to 0$,

$$\mu_f^{(-)} \underset{\sim}{=} -4k_1 c$$

In this case, the effective forward rate constant, k_f, is given by

$$k_f = 4k_1 \qquad\qquad\qquad 2.49$$

If, on the other hand, $k_{-1} \gg k_2''$ and k_{-2} is sufficiently small, then

$$\mu_f^{(-)} \cong - 4K_1 k_2'' c$$

and

$$k_f = 4K_1 k_2'' \qquad\qquad 2.50$$

In the first case, the reaction is rate limited by the bimolecular step; the first bond form

and the second is established instantaneously and essentially irreversibly within the time

the experiment. This is analogous to the pseudo steady state in <u>Michaelis–Menton</u> kineti

In the second "limit," equilibrium between free and singly bound antibody is rapidly esta

lished and occasionally perturbed, instantaneously, by formation of a second bond. It i

noteworthy that in both cases the relaxation time for the process $\mu_f^{(-)}$ is linearly propor-

tional to the antibody concentration in agreement with experimental observation.

Thus, there are a number of conditions which yield relatively simple expression

for the relaxation times of the forward and reverse rate constants, and it is pertinent to

whether any of these is relevant to the Hornick–Karush experiments. In attempting to pr

vide an answer, an important experimental finding to consider is that the equilibrium co

stant, K_{OBS}, obtained under equilibrium conditions, was found to equal k_f/k_r to within

factor of two. This places constraint on the possible interpretations of $\mu_f^{(-)}$. To be mo

specific, by definition,

$$K_{OBS} \cong \frac{\rho_1(\infty) + \rho_2(\infty)}{[\rho_0 - \rho_1(\infty) - \rho_2(\infty)] c}$$

where $\rho_1(\infty)$ and $\rho_2(\infty)$ are obtained by setting the right side of equation 2.47 equal to z

solving for $\xi(\infty)$ and $\rho_2(\infty)$ and using $\xi(\infty) = 2\rho_1(\infty)$. With the values thus obtained

$$K_{OBS} \cong 2K_1 c + 2K_1 c K_2'' \cong 2K_1 c K_2'' \qquad\qquad 2.51$$

Two ways in which one can have

$$\frac{k_f}{k_r} \cong 2K_1 K_2'' = \frac{2k_1 k_2''}{k_{-1} k_{-2}} \qquad\qquad 2.52$$

are

(A) $\quad k_r = 2k_{-2} \qquad$ (equation 2.40)

$$k_f = 4K_1 \, k_2'' \qquad \text{(equation 2.50)}$$

or

(B) $\qquad k_r = \dfrac{2k_{-1}}{K_2'} \qquad \text{(equation 2.41)}$

$$k_f = 4k_1 \qquad \text{(equation 2.49)}$$

For possibility (A), the reverse rate constant $k_r = 2k_{-2} = 3.3 \times 10^{-4} \text{ sec}^{-1}$. On the other hand, $k_{-1} = (1\text{-}100) \text{ sec}^{-1}$. Consequently there is a four to six order of magnitude difference between k_{-1} and k_{-2}. There is, however, no physical basis for such an enormous discrepancy. The main source of variation, if it exists, would be strain in the intramolecular complex and that would tend to make k_{-2} larger than k_{-1}. The magnitude of the difference is sufficient to rule out possibility (A) and I will therefore assume that k_r and k_f are given by equations 2.41 and 2.49, respectively. With this identification k_2 can be calculated. Using $K_1 = 6 \times 10^6 \text{ M}^{-1}$ and $k_1 = 3.7 \times 10^7 \text{ (m-sec)}^{-1}$ (Hornick and Karush, 1969), one has that $k_{-1} = \dfrac{k_1}{K_1} = 6 \text{ sec}^{-1}$. Therefore from equation 2.41 with $k_r = 3.3 \times 10^{-4}$ sec^{-1}, one finds that $K_2' = 3.5 \times 10^4$. Moreover, if $k_{-1} \underset{\sim}{\backsim} k_{-2}$, then $k_2' = 2.1 \times 10^5 \text{ sec}^{-1}$. This large value of k_2' is consistent with the conditions under which equations 2.41 and 2.49 were obtained.

There are a number of caveats which should accompany the acceptance of these results. The most important, and the one which I will consider briefly, is the extent to which equations 2.41 and 2.49 are consistent with the heterogeneity of the antibody population. Short of a complete analysis, which will not be undertaken here, a definitive statement cannot be made. However, it appears possible that the rate constants are representative of a relatively homogeneous subset of antibodies centered on the population mode.

This can be understood by considering the neutralization process. Since k_1 is relatively invariant (Pecht et al., 1972) and binding is essentially irreversible within the time of the experiment, then with excess antibody, the bound antibodies will be selected randomly from the entire population. As a result, the affinity distribution of those that are bound will reflect the distribution of the entire population. With this situation, the reactivation experiments should lead to a spuriously rapid relaxation time since low affinity antibodies will dissociate first. Consequently there would not be agreement between equ

brium and kinetic measurements. The fact that there is, probably means that selection i

not random, most likely because all antibody subpopulations are not in excess. This me

depletion of the relatively low concentration portions of the distribution, but with the ce

tral portion perhaps still remaining in excess. Moreover, since there are about 100 epit

per phage, most of the depletion will be the result of binding to epitopes other than the o

at the sensitive site. Inactivation may therefore be caused primarily by antibodies clust

in some interval about the mode. Still, this is not sufficient for agreement and there mus

be constraints on the shape of the distribution. In particular, it must not drop off too shar

ly in the vicinity of the mode, but at some distance, should fall rather precipitously.

If the results are, in fact, representative of a fairly homogeneous subpopulation

then they can be compared to the predictions of the theory developed in the first part of th

chapter. According to equation 2.15

$$\frac{K_2}{K_1} = \frac{3V}{4\pi <r>^3 N} \qquad 2.53$$

However, what is determined experimentally is K_2' and K_1. For a fully flexible model an

with 100 epitopes uniformly distributed over the phage surface, $K_2' = 12K_2$. Then from

equation 2.53, with $K_1 = 6 \times 10^6 M^{-1}$ (Hornick and Karush, 1969) and a standard state

1 mole/liter ($V = 10^3 cm^3$, $N = 6 \times 10^{23}$), one finds that $K_2' = 4.4 \times 10^4$, which compare

favorably with the experimental estimate of 3.5×10^4.

Even if the numerical value for K_2' is accurate (within a factor of three to four),

will not be generally applicable. The reason is that the parameters characterizing intra

molecular reaction reflect chemical reactivity and epitope spacing, as well as hinge flex-

ibility. A quantity of more fundamental interest is the proportionality constant between

K_2 and K_1. This is given by $3V/(4\pi <r>^3 N)$ which for a fully flexible hinge is 6×10^{-4}

M^{-1}. Thus

$$K_2 \simeq 6 \times 10^{-4} K_1 \qquad 2.54$$

The constant of proportionality depends only upon sequence in the heavy chain constant

region and should therefore be invariant for a particular antibody class. The concordar between theory and the Hornick-Karush experiments suggests that equation 2.54 may be reliable for IgG.

Before concluding this section, there is one more aspect of these experiments which I would like to consider. Hornick and Karush found that phage neutralization using monovalent antibody yields a value for K_1 about a factor of ten higher than what is determined from ligand binding experiments. A possible explanation for this difference is that the preparation of monovalent antibody used in the former experiments contains some bivalent impurity. (In ligand binding experiments, bivalent antibody can be used so the impurity problem does not exist.) Although this possibility was recognized and special precautions were taken in the preparative procedure, it is nevertheless pertinent to ask how sensitive the results are to percent impurity. Let c_1 be the concentration of monovalent antibody. Then the partition function for binding the sensitive site is

$$Z = 1 + 2K_2'' \, K_1 \, c + 2K_1 \, c + K_1 \, c_1 \qquad \text{2.55}$$

where 1 weights an unbound antibody and subsequent terms represent different types of binding to the sensitive portion of the phage. More specifically, the second term weights doubly bound antibody; the third, singly bound bivalent antibody; and the fourth, singly bound monovalent antibody. In terms of equation 2.55, the expressions for the concentration of the various species are

$$\rho_1 = \frac{2K_1 \, c \, \rho_0}{Z} \qquad \text{2.56}$$

$$\rho_2 = \frac{2K_1 \, K_2'' \, c \, \rho_0}{Z} \qquad \text{2.57}$$

$$\rho_3 = \frac{K_1 \, c_1 \, \rho_0}{Z} \qquad \text{2.58}$$

where ρ_1 and ρ_2 have the meanings introduced previously, and ρ_3 is the concentration of bound monovalent antibody. Then

$$K_{OBS} = \frac{(\text{concentration of neutralized phage})}{(\text{concentration of active phage}) \, c_1} \qquad \text{2.59}$$

which upon utilizing equations 2.56 – 2.58 becomes

$$K_{OBS} = K_1 \{1 + 2 \ (c/c_1) \ [1 + K_2'' \ ]\} \stackrel{\sim}{=} K_1 \ [1 + 2K_2'' \ c/c_1] \qquad\qquad 2.60$$

Thus the observed equilibrium constant is larger than the equilibrium constant for mono

valent binding by a factor of $1 + K_2''$ (c/c_1). This means that with $K_2'' = 10^4$, for exampl

an impurity of 0.1% would be sufficient to account for an order of magnitude difference be-

tween K_{OBS} and K_1.

An alternative explanation is that the result obtained by phage inactivation with

monovalent fragment is representative of a high affinity segment of the population (Horn

and Karush, 1972), whereas the results of ligand binding are more typical of an average

(see Chapter 3). More specifically, in the inactivation experiments, even if initial bind

is by a subpopulation, centered on the mode, dissociation is in this case very rapid and

there will be a reequilibration leading to preferential binding of the high affinity segmer

Thus even for a 100% pure monovalent population, heterogeneity could lead to a differen

between K_{OBS} and K_1.

2.5   Immunological Implications

The results of both the experimental and theoretical analyses presented here in

cate that for IgG, bivalence increases the equilibrium constant for binding antigen by al

four orders of magnitude. This means, in effect, that for equilibrium-controlled reactic

the concentration of bivalent antibodies needed to bind a given amount of antigen is abov

four orders of magnitude less than the required concentration of monovalent antibodies.

Moreover, the biosynthetic cost of this reduction, from an evolutionary point of view, is

minimal. The cross linking of identical heavy-light chain pairs to form bivalent antiboc

merely requires cysteine residues appropriately positioned for interchain disulfide bon

However, although multivalence and hinge flexibility confer a distinct binding advantag

the relation between this and biological activity is not always simple.

For some immunoglobulin functions the relation is clear. Binding of antibodies

bacteria and microbes facilitates phagocytosis so that any interactive advantage favors t

process (opsonization). In addition, the lytic ability of complement (a complex system

nonspecific factors in the blood and tissue fluids) is antibody dependent. Since cell lys

is mediated by interaction between complement and the Fc portion of bound immunoglob greater concentrations of bound antibodies lead to more efficient lysis and hence to more rapid destruction and elimination of foreign cells. Clearly, increasing valence is distinctly advantageous for performing these functions. From this point of view, multivalance is advantageous, and by implication, IgM would be preferable to IgG.

The observed presence of IgM early in the response is consistent with the above reasoning. It is advantageous for the host to make antibodies that are as efficient as possible since not enough time will have elapsed to produce large quantities of antibody. The situation, however, is more complicated since, for many antigens, there is a switch from IgM to IgG synthesis within about five or six days after immunization. This suggests that simply maximizing valence may not be optimal for species survival. The other factors involved, however, are not clear. Possibly the opsonic activity of IgM is so potent and its complement-fixing capability so great, that large concentrations over prolonged periods of time are destructive to self-components that cross react (weakly) with the foreign antigen. This being so, as antigen is removed, the need for the most "efficient" antibody would no longer be as acute, and a switch to IgG synthesis would become advantageous.

In addition to its role as antibody in destroying and eliminating antigen, immunoglobulin is present on lymphocyte membranes. In particular, B cells appear to have bivalent IgM and IgD receptors (Vitetta and Uhr, 1975). It is therefore pertinent to ask what role multivalence plays here. Since detailed information on B cell activation in relation to antigen receptor interactions is just starting to become available, remarks in this area must necessarily be speculative. However, there are some pertinent observations.

First, although monovalent receptors can to a certain extent be cross linked by antigen, the formation of an antigen receptor lattice involving more than one antigen is not possible. However, the biological significance of this distinction is difficult to assess at present. Although there is some evidence that cross linking is necessary for activation (Chapter 8), there appears to be no direct evidence bearing on the requirements of lattice formation.

In addition to making lattice formation possible, bivalence allows intramolecular bonding and this too, as was shown, provides an enormous increase in stability. However, since there is competition between intramolecular reaction and cross linking, it is pertinent to ask what the relation between the two might be.

There are four variables of particular interest in determining the competition be
tween cross linking and intramolecular reaction. First, as epitope density on the antige
decreases, it becomes progressively more difficult to establish a second bond by either
cross linking or intramolecular reaction. In fact, according to equations 2.20 and 2.21,
a high uniform determinant density, there is no preference. However, if the antigen is la
and the determinant density very low, intramolecular bonding may become impossible, wh
cross links may still be established with relative ease. Consequently in this limit, reducti
in determinant density are expected to lead preferentially to cross linking, although the
likelihood of both will be reduced.

Second, as the membrane becomes more fluid, the collision rate between free re
ceptors and membrane-bound antigens increases and this too favors cross linking. The
same is true of an increase in receptor concentration.

Finally, it is clear from equation 2.22 that increasing $<r>$, i.e., increasing hin
flexibility also favors cross linking at the expense of intramolecular reaction. Thus, if
cross linking is required for B cell triggering, these conditions would tend to favor acti
vation, whereas the reverse conditions which shift the reaction pathways toward intramolecular reaction would tend to militate against it. The question then is whether too m
intramolecular reactions merely passively block activation or do something more. This
be discussed elsewhere (DeLisi, 1976) since conjecture here would lead too far from the
main theme of this chapter. The main idea here was to identify the parameters affecting
strength of the interaction and, insofar as possible, to make quantitative estimates of the
This allows some insight into which of several reaction pathways may be followed (intra
molecular reaction versus dissociation, cross linking versus intramolecular, etc.) and
into the potential biological significance of various structural features. It should be evi
that a great deal of effort still needs to be expended in this area, and a reasonably comp
understanding will only be achieved, if at all, by combined experimental and mathemati
methods.

3

COMBINING SITE HETEROGENEITY. THE FREE
ENERGY DISTRIBUTION FUNCTION

3.1   Statement of the Problem

According to the theory of clonal selection, antigen may bind to and stimulate proliferation and differentiation of a number of distinct B cells. Although each of the resulting
clones produces homogeneous antibody, the entire population will be heterogeneous because
of heterogeneity in the B cell population. The combining site diversity thus generated in
the antiserum will be reflected in heterogeneity of free energies and equilibrium constants
(affinities) characterizing its interaction with antigen. A complete thermodynamic description of binding therefore requires knowledge of a distribution of free energies.

In trying to understand how the distribution might be specified, it is useful to begin by considering the sources of antiserum heterogeneity. As indicated in Chapter 1, even
a single determinant will elicit a heterogeneous response. In addition, however, natural
antigen generally possesses a variety of different determinants, and since each elicits its
own set of antibodies, antiserum heterogeneity will be a weighted sum of heterogeneous
responses to each different determinant.

Fortunately, it is easy to design experiments which obviate having to consider
heterogeneity arising from determinant diversity by using a hapten-carrier complex as immunogen. Antibodies will then be produced against the hapten, as well as determinants on
the carrier, but if only an antiserum-hapten mixture is studied, only the antibodies directed
against hapten will enter the results, the anti-carrier antibodies being in effect inert with
respect to the hapten.

A particularly useful experimental procedure for studying binding properties of
antiserum is <u>equilibrium dialysis</u> (Eisen, 1964). The method yields the concentration of
bound antibody sites as a function of free hapten concentration. The problem of interest is
to determine the distribution of binding free energies for hapten from experimental data of
this sort. It should be noticed that in this case, intramolecular reaction is not possible
since the hapten is approximately the size of the antibody combining site.

The problem can be formulated as follows: Let $n(G)dG$ be the fraction of antibody combining sites whose free energy of interaction with hapten is between G and $G+dG$, and let $r(G)$ be the fraction of those sites that are bound. Then if H is the free hapten concentration, by mass action

$$K_1 = \frac{r(G)}{[1 - r(G)]H} \qquad 3.1$$

or

$$r(G) = \frac{K_1 H}{1 + K_1 H} \qquad 3.2$$

K_1 is, as defined previously, the antibody combining site–hapten equilibrium constant. Defining $f_0(H,G)dG$ as the fraction of the entire population of antibody sites that has reacted with free energy between G and $G+dG$ when the hapten concentration is H, it is evident that

$$f_0(H,G)dG = r(G)n(G)dG = \frac{K_1 H n(G) dG}{1 + K_1 H} \qquad 3.3$$

or

$$f(H) = \int_{-\infty}^{\infty} f_0(H,G) \, dG = \int_{-\infty}^{\infty} \frac{e^{G/RT} H n(G) \, dG}{1 + e^{G/RT} H} \qquad 3.4$$

where $f(H)$ is a function obtained by fitting experimental data. The derivation of course assumes that the free energy distribution is sufficiently dense and that binding is independent of the state of neighboring antibody sites.

3.2 Inversion of the Binding Curve

The problem is to solve equation 3.4 for $n(G)$, the free energy distribution function. Introducing the exponential substitution

$$\xi = \exp(G/RT) \qquad 3.5$$

equation 3.4 becomes

$$f\left(\frac{1}{\eta}\right) = RT \int_0^{\infty} \frac{\nu(\xi) \, d\xi}{\xi + \eta} \qquad 3.6$$

where

$$\eta = \frac{1}{H} \qquad\qquad 3.7$$

and

$$\nu\ (\xi) = n\ (RT\ \ln\ \xi) \qquad\qquad 3.8$$

Equation 3.6 is a singular integral equation of the first kind with the symmetric kernal $(\xi + \eta)^{-1}$. Provided the experimental data can be fit by a function which is analytic and single valued in the complex plane cut along the negative real axis, f is the Stiljes transform of ν. In particular (Erdélyi, 1954),

$$\nu\ (\xi) = \frac{f(\xi e^{-\pi i})\ -f(\xi e^{\pi i}\ )}{2\pi i} \qquad\qquad 3.9$$

In addition to being analytic, f must meet the physical requirements that

$$f\ (0) = 0$$

and $f(\infty) = 1$

This last condition is equivalent to requiring that

$$\int_{-\infty}^{\infty} n\ (G)\ dG = 1$$

Sips (1948) has assumed that for the binding of gas molecules to a heterogeneous catalytic surface, the left side of equation 3.6 can be represented by a modified Freundlich isotherm; i.e., by

$$f(H) = \frac{AH^a}{1 + AH^a} \qquad 0 < a \le 1 \qquad\qquad 3.10$$

where a and A are constants which are to be determined. Equation 3.10 was chosen since it has the desired saturation property at large H, viz., the fraction of bound antibody sites approaches 1 as the number of hapten molecules available increases. At small H, however, it is not evident that the form is reasonable since the fraction of bound sites might be expected to be a linear function of H.

Using equation 3.10 for $f(H)$, one finds that the free energy distribution function is given by (Sips, 1948)

$$n(G) = \frac{1}{\pi RT} \; \frac{A \exp(-aG/RT) \sin\pi a}{1 + 2A \exp(-aG/RT) \cos\pi a + A^2 \exp(-2aG/RT)} \qquad 3.11$$

Because of the unrealistic behavior of equation 3.10 at very low H, and because the range of integration extends over negative G, the result can at best be an approximation valid over a finite free energy interval.

The distribution has the following properties:

1. It is a probability distribution, i.e.,

$$\int_{-\infty}^{\infty} n(G) \, dG = 1 \qquad 3.12$$

2. The distribution is symmetric. This follows since there is a single maximum whose position is at the mean of the distribution. Setting the derivative of equation 3.11 equal to zero, it is readily seen that the position of the maximum is at

$$G_m = \frac{RT}{a} \ln A \qquad\qquad a \neq 0 \qquad 3.13$$

Moreover, by a straightforward integration

$$\bar{G} = \int_{-\infty}^{\infty} G \, n(G) \, dG = \frac{RT}{a} \ln A \qquad 3.14$$

Using equation 3.13, the distribution, 3.11, can be written as

$$n(G_m-G) = \frac{1}{\pi RT} \; \frac{\exp[a(G_m-G)/RT] \sin\pi a}{1 + (2\cos\pi a) \exp[a(G_m-G)/RT] + \exp[2a(G_m-G)/RT]} \qquad 3.15$$

3. The distribution represents a homogeneous antibody population as a approac 1. The two relevant properties are

(a) $\displaystyle \lim_{a \to 1} n(G-G_m) = 0 \qquad G \neq G_m$

(b) $\quad \lim\limits_{a \to 1} \displaystyle\int_{-\infty}^{\infty} n(G - G_m)\, dG = 1$

Property (a) is evident from equation 3.15. Property (b) follows by a simple integration. Consequently, in this limit there is only one species present.

The parameter, a, is in fact related to the width of the distribution. More specifically, the half width at half maximum, $G_{\frac{1}{2}}$, is given by

$$G_{\frac{1}{2}} = \frac{RT}{a} \ \ln\left\{2 + \cos\,(\pi a) - \sqrt{(2 + \cos \pi a)^2 - 1}\,\right\}$$

so that there is an approximate reciprocal relation between the width and a.

The width and amplitude of the distribution are therefore characterized by the two parameters, a and A. In order to complete the description, they must be identified with experimentally measureable quantities. Substituting A from equation 3.13 into equation 3.10, one finds that

$$f(H) = \frac{[H \exp\,(G_m/RT)]^a}{1 + [H \exp(G_m/RT)]^a} \qquad\qquad 3.16$$

This can be written in the form

$$\ln\left(\frac{f(H)}{1 - f(H)}\right) = a \ln\,(H) + \frac{aG_m}{RT} \qquad\qquad 3.17$$

Equation 3.17 predicts that a plot of $\ln\left[f/(1-f)\right]$ against $\ln H$ will be linear, and will thus specify both a and G_m.

This procedure is unfortunately not very general since there will be circumstances for which a plot conforming to equation 3.17 will not produce a straight line (see below). The simplest extension is to write f(H) as a weighted sum of two terms of the form of equation 3.10, viz.,

$$f(H) = \frac{w_1\, A_1\, H^{a_1}}{1 + A_1\, H^{a_1}} + \frac{w_2\, A_2\, H^{a_2}}{1 + A_2\, H^{a_2}} \qquad\qquad 3.18$$

with

$$w_1 + w_2 = 1$$

This is equivalent to assuming that the free energy distribution $n(G)$ can be represented a

$$n(G) = w_1 n_1(G) + (1 - w_1) n_2(G) \qquad 3.19$$

where $n_1(G)$ and $n_2(G)$ are distributions of the form 3.15.

A completely different procedure can be developed by beginning with the observa tion that the limits of integration in equation 3.5 should actually be finite. In particular, consider

$$f(e^{-x/RT}) = \int_0^x \frac{e^{(G-x)/RT} n(G) \, dG}{1 + e^{(G-x)/RT}} \qquad 3.20$$

where $e^{x/RT}$ has been substituted for H. The distribution found from this equation will b inaccurate principally because the upper limit extends only to x. As a result, only equili brium constants exceeding the value $KH = 1$ will be included. This deficiency can be cor- rected by replacing the upper limit by αx ($\alpha > 1$) where α is chosen sufficiently large so that all equilibrium constants are included in the integration. Then,

$$f(e^{x/RT}) = \int_0^{\alpha x} \frac{n(G) \, dG}{1 + e^{(x-G)/RT}} \qquad 3.21$$

Introducing

$$x = \frac{RTy}{\alpha}, \quad G = \frac{RTt}{\alpha} \qquad 3.22$$

Equation 3.21 can be written as

$$\hat{f}(y) = \int_0^y \hat{n}(t) \, \hat{K}(y - t) \, dt \qquad 3.23$$

where

$$\hat{f}(y) = \frac{\alpha}{RT} f(e^{y/\alpha}) \qquad 3.24$$

$$\hat{n}(t) \equiv n(RTt/\alpha) \qquad \qquad 3.25$$

$$\hat{K}(y-t) \equiv [1 + \exp(\frac{y}{\alpha} - \frac{t}{\alpha})]^{-1} \qquad \qquad 3.26$$

The formal solution for the affinity distribution can now be obtained using Laplace transforms

$$\hat{n} = \frac{1}{2\pi i} \int_{c-i\infty}^{c+i\infty} \frac{F(y)}{G(y)} dy \qquad \qquad 3.27$$

where F and G are the Laplace transforms of $\tilde{f}$ and $\tilde{K}$.

3.3   Intramolecular Reactions

In experiments with free hapten and antibody, intramolecular reaction does not occur because the ligand is approximately the size of the antibody combining site. However, the experimental procedure can easily be modified so that intramolecular bonding occurs, by using ligand covalently bound to a large molecule or cell. Then an experiment which measures the fraction of bound sites as a function of free hapten concentration can be used to obtain information about the intramolecular equilibrium constant.

In such an experiment let $f_1(H)$ and $r_1(G)$ denote, in analogy to $f(H)$ and $r(G)$, the generator and kernal of the transform. Then

$$f_1(H) = \int_{-\infty}^{\infty} r_1(G)\, n(G)\, dG \qquad \qquad 3.28$$

n(G) has precisely the same meaning as in equation 3.4 since it depends only on the immunization and not on the dialysis experiment. That is to say, n(G) is the fraction of antibodies having free energy G for hapten and this depends only on the response of the animal. $r_1(G)$ must, of course, be different from r(G) because these functions represent the fraction of a subpopulation which is bound and that will depend upon the types of interactions which are possible. This difference will be reflected in a difference between $f_1(H)$ and $f(H)$.

With n(G) regarded as known from ligand binding experiments, the main problem is to obtain $r_1(G)$ as a function of K_2. Although this can be done quite generally (DeLisi and Perelson, 1976), experimental conditions can be set so that a relatively simple approximate expression is adequate.

In particular, one can carry out experiments in excess H, so that little cross link occurs (see Chapter 8). In this case, the main complexes present will be those shown in Figure 3.1.

It is evident that r_1 (G) can be written as a sum of terms over the antibody states shown in Figure 3.1, the ith term being the probability that an antibody is in state i, mult plied by the fraction of its sites that are bound when it is in that state. The probabilities are easily written in terms of the partition function for these states.

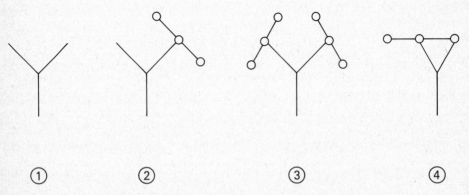

Figure 3.1. An antibody may be free (state ①), singly (state ②) or doubly bound. In the last case one (state ④) or two (state ③) antigens may be involved.

According to the rules of statistical thermodynamics (see for example, Hill, 1960 p. 145), the weights for states ① - ④ are, respectively, 1, $2K_1H$, $(K_1H)^2$ and $2K_1H\ K_2'$ K_2' is the equilibrium constant for intramolecular reaction that was introduced in Chapter The partition function Z is, by definition, just the sum of these.

$$Z = 1 + 2K_1H + (K_1H)^2 + 2K_2'\ KH \qquad\qquad 3.29$$

Therefore the fraction of bound sites is given by

$$r_1\ (G) = \frac{K_1H\ (1 + K_1\ H + 2K_2')}{Z} \qquad\qquad 3.30$$

When $K_2' = 0$, this reduces to equation 3.3 as it should.

Proceeding as in the derivation of equation 3.5, one has that

$$f_1(H) = H \int \frac{e^{G/RT}}{Z} \; (1 + He^{G/RT} + 2\gamma\, e^{G/RT})\, n(G)\, dG \qquad\qquad 3.31$$

where $\gamma = K_1/K_2'$. For fixed H and with $n(G)$ known, 3.31 is nonlinear algebraic equation for γ, with the integration taken over some reasonable interval (e.g., $5 \leq G/RT \leq 25$).

3.4 Immunological Implications

Extensive binding studies have been carried out by Werblin and Siskind (1972) using anti-DNP antibody obtained from rabbit at various times after immunization. Data from serum taken at days 7 and 42 plotted according to equation 3.17 look something like this.

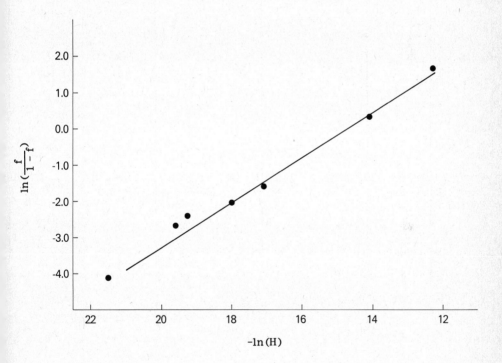

Figure 3.2. The results of binding experiments using antiserum taken seven days after immunization.

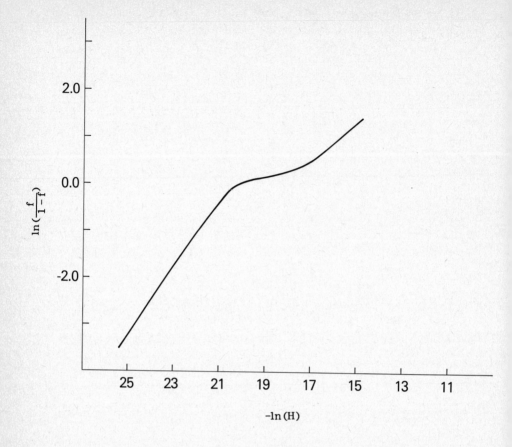

Figure 3.3. The results of binding experiments using antiserum taken 42 days after immunization.

The seven-day data fall pretty nearly on a straight line so the use of equation 3.16 for $f(H)$ seems reasonable. The resulting distribution (dashed line, Figure 3.4) has G_m = 7.9 Kcal and a = 0.58.

For the 42-day data, there is a clear break in the curve. Consequently, it is no longer appropriate to use equation 3.17 and the distribution will not be given by 3.15. The simplest extension of the above procedure is to represent the data as a weighted sum of two modified Freundlich isotherms. The viability of this approach is suggested by the linearity of the curves at high and low hapten concentrations. A straight line fit in each of these regions yields the parameters a_1, A_1, a_2, and A_2. The resulting distribution is shown in Figure 3.4 (solid line).

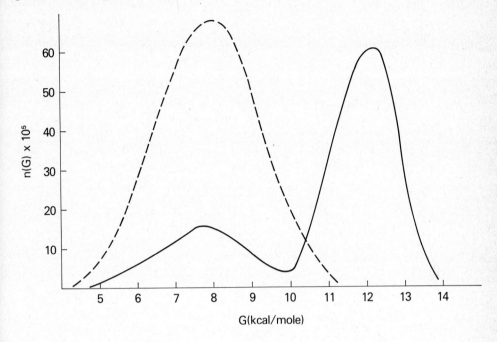

Figure 3.4. Free energy distributions obtained from data taken seven (- - -) and 42 days (——) after immunization.

A more sophisticated numerical procedure has been used by Werblin and Siskind (1972). They wrote equation 3.4 in the form

$$f(H) = \sum_{i=1}^{N} \frac{n(K_i) K_i H}{1 + K_i H}$$

where K_i is the affinity of the ith antibody species and N the (unknown) number of antibody subpopulations, and used a nonlinear optimization procedure to determine $n(K_i)$ for fixed N. The distribution thus obtained was used to calculate the binding curve, and the sum of the squares of the deviation between the theoretical and experimental points used as a measure of the "goodness" of the distribution. Their results are in substantial agreement with those presented here.

The antibody affinity distribution is expected to reflect a number of cellular char
acteristics of the immune system. First, it should be a function of the B cell receptor affi
distribution. This distribution is not yet known experimentally, although a reasonable gue
is that it increases monotonically as the free energy decreases (i.e., as the free energy
becomes less favorable). This is in accord with the idea that a favorable interaction impl
some sort of structural compatibility between determinant and combining site. As the requirement for precise compatibility is relaxed, there will be a wider range of receptor sit
with which the determinant can interact, although less favorably. On the basis of this re
soning, one would predict that a distribution of the sort shown in Figure 3.5 (-O-) is expected. Consequently, if the antibody affinity distribution reflected only the affinity of
receptors on precursors, $n(G)$ would be a monotonically decreasing function of free ener

Since the results shown in Figure 3.4 indicate that the distribution is peaked, th
must be other sources making important contributions. The most obvious is that the prob
bility of B cell activation (i.e., triggering differentiation and proliferation) is free energ
dependent. In particular, a peaked distribution can arise if the probability of triggering
increases as the binding free energy increases as indicated by the solid line in Figure 3.
In fact, such a relationship is expected according to clonal selection since binding is a
necessary requirement for triggering, and the chance of binding clearly increases with
increasing energy.

The triggering curve in Figure 3.5 has been drawn with a maximum for theoretic
reasons which will be discussed in Chapter 8 and which are related to the distinction betv
binding reactions and triggering reactions. The main point here is the conjecture that th
observed peak in the former can arise as a juxtaposition of two cellular characteristics wh
depend in reciprocal ways upon the interaction free energy over some range in free energi

Although the above argument assumes that the peak arises as the result of an advan
tage which high affinity cells have in capturing antigen, early in the response this advantage cannot be very great, since there will be so much antigen present that even those of low
affinity will have a substantial fraction of their receptors bound. However, even a small
advantage will be amplified as the result of clonal expansion. For example, if two subpop
ulations are compared, one having a ten percent advantage over the other with regard to ,
say, speed of proliferation, then over 30 generations the differential amplification factor

will be $(1.1)^{30}$ = 17.4. Actually, what is important is not just proliferation, but some combination of proliferation and differentiation that leads over several generations to differential amplification in the total quantity of antibody produced by a clone. In fact, it is possible that such differential amplification through clonal expansion is so important that it almost entirely obliterates the effect of the precursor distribution. If this is so, then n(G) provides considerable insight into B cell triggering characteristics, and its biological significance would be even greater than had previously been expected.

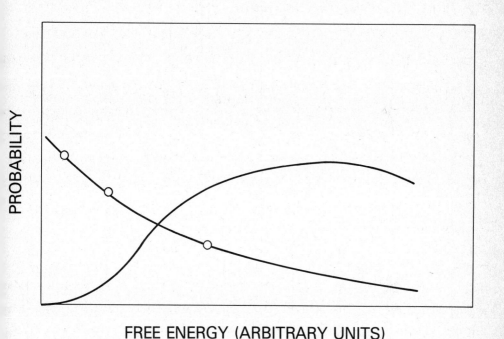

Figure 3.5. The ordinate is the probability of finding a precursor with receptors having a given free energy for antigen (-o-), or the probability of triggering (——).

Aside from the peak in the distribution, the results in Figure 3.4 indicate an increase in affinity as time after immunization increases. This can be interpreted in two ways. First, as time increases, there may be preferential stimulation of high affinity cells and

consequently preferential production of high affinity antibodies. Alternatively, the affin
of antibodies being synthesized may be independent of time after immunization. In this
case the large antigen concentration present early in the response would bind high affin:
antibody, resulting in its selective removal. Some insight into which possibility is corr
is provided by the experiments of Steiner and Eisen (1967). By cultivating lymphoid ce'
from immunized rabbits and measuring the affinity of antibody secreted, they showed the
early immune cells synthesize lower affinity antibody than those produced later in the re
sponse. This implies selective stimulation of cells having high affinity receptors as time
after immunization increases.

Pressure for selection probably arises when there is not enough free antigen av
able to bind all cellular receptors. In such a situation there will be preferential binding
to high affinity subpopulations and this advantage should continue to increase as the fre
concentration decreases.

This general behavior is consistent with the results shown in Figure 3.4. The
mean energy at day 7 is 7.9 Kcal whereas at day 42 it is 10.3 Kcal. Moreover, the form
the affinity distribution is also changing. At day 7 it is approximately symmetric, wher
at day 42 it has developed a long low-affinity tail. This sort of shift in the distribution i
what one would expect on the basis of a concentration-dependent selection mechanism:
high affinity B cells are being preferentially, though not exclusively, stimulated as anti
is depleted.

This selection hypothesis is also consistent with a variety of other thermodynan
data which will not be reviewed here. However, one example of particular interest is t
effect of administering, along with antigen, specifically reactive antibody. As one migh
expect, this has a suppressive effect on the response, presumably because the amount o
antigen available for binding B cells is reduced as the result of complexation with antib
(Goidel et al., 1968). More importantly, however, the low affinity cells are preferentia
suppressed (Harel et al., 1970). This is precisely what one would expect since those c
compete least effectively.

4

COMBINING SITE HETEROGENEITY.
IMMUNODIFFUSION

4.1   General Remarks

There are many in vitro, as well as in vivo, circumstances in which chemical reaction of antigen with antibody is coupled to the presence of time varying concentration
gradients of one or both reactants. In this chapter I will consider a simple though immunologically interesting example of this, and show how it can be used to determine information about antiserum characteristics.

In the situation of interest, antigen (antiserum) is placed in a small circular well
which is located in a uniform field of an antiserum (antigen)-agar gel mixture. At a micro-
scopic level, the gel is highly porous, having the appearance of an extended matrix. If a
molecule (antigen or antibody) is small compared to the average pore size, its movement
will be relatively unimpeded and it will diffuse to a good approximation as though it were
in water. If the reacting unit is large or rigidly attached to a large object, diffusion may
be hindered or entirely prevented.

Antibody molecules are generally small compared to average pore size, so their
motion is relatively unrestricted by the gel. Antigenic determinants, on the other hand,
are often part of large objects such as red blood cells and in such cases they are immobile.
Two situations may therefore be distinguished, depending upon the relationship between
antigen size and gel pore size.

One type of experiment typically involves radial diffusion of antigen, over a period
of about 20 hours, from a circular well into a gel containing specific antiserum (Mancini
et al., 1965; Williams and Chase, 1971). In this case the antigen must, of course, be relatively small. The first step in a reaction with antibody is univalent attachment. This
may be followed by a number of events, including intramolecular reaction (both sites bound
to different determinants on the same antigen), cross linking (each site bound to a different
antigen), or dissociation.

If the initial concentrations of antigen and antibody are chosen appropriately, and the antibody is sufficiently reactive, cross linking will be possible and large branched a gen-antibody networks may form. As these aggregates grow, they diffuse less readily a finally precipitate from solution.

Procedures of this sort provide a means of testing for specific reactivity of antig to a particular serum or vice versa. In addition, Mancini et al. have observed a linear relationship between precipitate area and total antigen concentration when the reaction has gone to completion, thus providing a basis for the quantitative use of the method. A mathematical theory of immunodiffusion leading to precipitation has also been developed by Aladjem and Palmiter (1965) and applied to the estimation of diffusion coefficients.

The utility of this method can be readily extended by applying a homogeneous unidirectional electric field parallel to the plane of the agar layer (Laurell, 1966). If th system is buffered to maintain a pH near the isoelectric point (i.e., zero net charge) of most antibodies, primarily antigen will be transported. In addition, if there are two or more types of antigen in the well, they will generally be transported at different velocit by the field. The result is a series of precipitate bands.

The analytical sensitivity of the method may be increased by applying a second at right angles to the first, after initial separation. Moreover, both methods can to som extent be used quantitatively, since after antigen has left the well, various geometrical features of the precipitate pattern can be empirically related to antigen concentration. the exception of some numerical calculations (Cann, 1975), there has been relatively lit mathematical work in this area.

4.2   A Simple Diffusion-Reaction Model for Plaque Formation

4.2.1   Basic Concepts

The procedures just described involve diffusion or transport of antigen (antise from a well into a surrounding medium where it interacts with antiserum (antigen). Pr pitation is the observable consequence. An alternative method for detecting reactions i hemolysis (red blood cell lysis). In this case the reactants consist of RBCs distributed uniformly throughout the agar medium, and their specific antiserum which is initially i the well (Weiler et al., 1965; Hiramoto et al., 1970). The addition of complement to the mixture causes, by some poorly understood mechanism, membrane lesions and consequ loss of hemoglobin in those RBCs having a sufficient amount of bound antibody. The result is a clear circular plaque surrounding the well. This is the type of experiment that will be modeled.

The RBC membrane is highly heterogeneous with respect to its antigenic properties, i.e., it possesses many different types of determinants, each occurring with some characteristic frequency and each capable of stimulating antibody production. This heterogeneity complicates the analysis. However, if a hapten carrier complex is used as immunogen, and the RBCs in the assay are modified by covalently coupling the hapten to their surface membranes, the antihapten antibodies see a homogeneous set of sites (hapten) on the RBC.

The size of a plaque at a particular time will depend upon the initial concentration of antiserum, as is commonly recognized (Weiler et al., 1965; Hiramoto et al., 1970), but it will also depend upon other quantities, including reaction rate constants. Our purpose is to derive expressions relating measureable quantities to physical chemical parameters and to apply the results to the analysis of experimental data on IgM antibodies.

Consider the diffusion of antibodies from a circular well of radius a, into a two-dimensional gel containing haptenated RBCs with which the antibodies interact specifically and reversibly. Let $\vec{r}$ and t measure, respectively, the distance from the center of the well and the length of time which elapsed since incubation began. As time increases, antibodies diffuse into the gel where they interact with RBC-bound hapten. Since antibodies have two or more combining sites, univalent interaction between a single site and an RBC epitope may be followed by an intramolecular step leading to multisite attachment. However, intramolecular reactions will be unlikely if the epitopes are distributed sparsely enough so that the mean distance between them cannot be spanned by the combining sites on a single antibody. This is the situation that will be considered.

4.2.2   The Equations

The equations of the model are those proposed by DeLisi and Bell (1974) and subsequently used by Goldstein, DeLisi and Abate (1975) to describe diffusion from a circular well. Let i be a running index distinguishing antibodies of different affinities, and let $C_{0i}(r,t)$ and $C_{1i}(r,t)$ be the concentrations of diffusing and bound antibodies of type i.

Then for $r > a$

$$\frac{\partial C_{0i}}{\partial t} = D_0 \nabla^2 C_{0i} - \frac{\partial C_{1i}}{\partial t} \qquad\qquad 4.1$$

and

$$\frac{\partial C_{1i}}{\partial t} = 2k_1 C_{0i} \rho - k_{-1i} C_{1i} \qquad\qquad 4.2$$

where ρ is the concentration of free epitopes, k_{-1i} the epitope-combining site dissociation rate constant for the ith species, and D_0 the diffusion coefficient. Equations 4.1 and 4.2 are to be solved subject to the initial condition that there is no antibody outside the well $t = 0$, and subject to the boundary condition that the antibody concentration vanishes isotropically at large distances from the center of the well.

$$C_{0i}(r,0) = 0 \qquad\qquad\qquad r > a$$

$$C_{1i}(r,0) = 0 \qquad\qquad\qquad r \geq 0$$

$$\lim_{r \to \infty} C_{0i}(r,t) = 0$$

For $r < a$, there is only antiserum and therefore no chemical reaction. Consequently

$$\frac{\partial C_{0i}}{\partial t} = D_1 \nabla^2 C_{0i} \qquad\qquad r < a \qquad\qquad 4.3$$

D_1 being the diffusion coefficient inside the well. This is to be solved subject to the condition that the ith species is at concentration C_i' at $t = 0$. The two solutions are matched by imposing a continuity requirement on the diffusion current at $r = a$.

According to equation 4.1, the concentration of antibody at t,r changes as the result of diffusion (the first term on the right) and chemical reaction with RBC epitope (the second term on the right). Equation 4.2 describes a simple bimolecular chemical reaction. Equations 4.1 - 4.3 must, in general, be solved numerically. However, for the experiments under consideration, some simplifications arise. One is that the epitope concentration is, under a wide range of conditions, in great excess over the antibody concentration. This is true locally as well as globally and allows the linearization of equation 4.2. Replacing the free epitope concentration by the total epitope concentration, ρ_0, one has that

$$\frac{\partial C_{1i}}{\partial t} = 2k_1 C_{0i} \rho_0 - k_{-1i} C_{1i} \qquad \qquad 4.4$$

It should also be noticed that by this same approximation, the equations describing the diffusion of different species are uncoupled from one another.

A further simplification arises when the time scale for chemical reaction is considered. Since typical relaxation times for dissociation are $\lesssim 1$ sec, changes resulting from reaction occur rapidly compared to those due to diffusion, and local equilibrium may be assumed. Thus

$$C_1 = 2K_1 \rho_0 C_0 \qquad \qquad 4.5$$

and

$$\frac{\partial C_1}{\partial t} = 2K_1 \rho_0 \frac{\partial C_0}{\partial t} \qquad \qquad 4.6$$

K_1 being the combining site-epitope equilibrium constant defined in Chapter 2. The subscript i has been dropped for notational simplicity. Substituting equation 4.5 into equation 4.1

$$\frac{\partial C_0}{\partial t} = D_2 \nabla^2 C_0 \qquad \qquad r > a \qquad \qquad 4.7$$

where

$$D_2 \equiv \frac{D_0}{1 + 2K_1 \rho_0} \qquad \qquad 4.8$$

Although equation 4.7 was derived under conditions of single site antibody attachment, it is actually applicable under somewhat more general conditions. In fact, it can be shown that even allowing intramolecular associations and dissociations equation 4.7 will become accurate as the time of the experiment becomes long, relative to the reciprocal of the

reverse rate constants (Section 5.4). In that case, of course, K_1 will be a complicated fu[...]
tion of the four rate constants for the elementary steps.

4.2.3 The Plaque Radius

The quantity of interest is $N(r,t)$, the number of antibodies bound per RBC, sin[...]
this determines the probability of lysis. It is generally assumed that only when $N(r,t)$ i[...]
larger than some minimum value, N^*, will the cell lyse (DeLisi and Bell, 1974; Jerne et [...]
1974). This means the plaque radius, r_p, is the distance at which there are, on the avera[...]
N^* antibodies bound per RBC. However, there are two complications which must be con[...]
dered. First, the plaque radius will not be sharp since at distances somewhat greater th[...]
r_p, the number of antibodies bound may not be so small as to make lysis imperceptible. [...]
will the number bound at distances slightly less than r_p be sufficient for complete lysis. [...]
addition, if N^* is small, as will undoubtedly be the case for IgM, fluctuations from its mean[...]
may have to be considered.

To express the first of these effects more formally, let $N(r_p \pm \delta/2)$ be the numb[...]
of antibodies bound per RBC at $r_p \pm \delta/2$. Then

$$N(r_p \pm \delta/2) = N^* \overline{\mp} \varepsilon \qquad\qquad 4.9$$

For small deviations from r_p,

$$N(r_p \pm \delta/2) = N(r_p) \pm \left.\frac{\partial N}{\partial r}\right|_{r_p} \frac{\delta}{2} \qquad\qquad 4.10$$

$$\equiv N^* \pm \left.\frac{\partial N}{\partial r}\right|_{r_p} \frac{\delta}{2}$$

Thus the spread in the number of bound antibodies per RBC over a distance δ centered c[...]
r_p is:

$$N(r_p + \frac{\delta}{2}) - N(r_p - \frac{\delta}{2}) \equiv 2\varepsilon \stackrel{\sim}{=} \left.\frac{\partial N}{\partial r}\right|_{r_p} \delta$$

or

$$\delta = \frac{2\varepsilon}{\left.\frac{\partial N}{\partial r}\right|_{r_p}} \qquad\qquad 4.11$$

Therefore δ varies inversely as the gradient at r_p. With time fixed, one generally expects smaller gradients at larger distances. Consequently, there will be greater uncertainty in the radii of large plaques.

Stochastic effects on the number of antibodies bound at a fixed distance may be discussed approximately by assuming that complement is added at a time when the concentration of diffusing antibodies is near its steady state value. Let P_N be the probability that a cell has N bound antibodies, and let $f_N(t)$ be the probability that a cell with N bound antibodies acted on by complement for time t is lysed. Then the fraction of lysed cells, P_L, is

$$P_L = \sum_{N=0}^{\infty} P_N f_N(t) \qquad\qquad 4.12$$

If the rate of lysis of cells with N antibodies is proportional to the number of unlysed cells with N antibodies, then

$$f_N(t) = 1 - e^{-\lambda N t} \qquad\qquad 4.13$$

where λ is the rate constant for cell lysis by complement when only a single antibody is bound. In addition, if N is poisson distributed with mean N^*, i.e., if

$$P_N = \frac{N^{*N} e^{-N^*}}{N!} \qquad\qquad 4.14$$

then equation 4.12 becomes

$$P_L = \sum_{N=0}^{\infty} \frac{N^{*N} e^{-N^*}}{N!} - \sum_{N=0}^{\infty} \frac{(N^* e^{-\lambda t})^N e^{-N^*}}{N!}$$

or (Jerne et al., 1974)

$$P_L = 1 - \exp[-N^*(1 - e^{-\lambda t})] \qquad\qquad 4.15$$

This defines the plaque boundary as a function of N^* since P_L will have a fixed value there. Jerne assumes $P_L = 2/3$, $t = 10^3$ and $\lambda = (10^{-3} - 10^{-4})$ sec^{-1} to obtain $N^* = (1 - 10)$ antibodies/RBC. In the theories which follow, I will assume that at $r = r_p$, $N^* = 5$.

With N^* fixed in this manner, the plaque radius can be determined by solving t
equations for $N(r,t)$. From equation 4.5, with E the number of epitopes per RBC,

$$N(r,t) = 2 \sum_i E K_{1i} C_{0i} (r, t) \qquad\qquad 4.16$$

so that the plaque radius is obtained by solving

$$N^* = 2 \sum_i E K_{1i} C_{0i} (r_p, t) \qquad\qquad 4.17$$

for r_p. The problem, therefore, is to find $C_{0i}(r,t)$.

4.3   The Solution

A formally identical problem has been investigated in connection with the theor
of heat conduction. Although uniformly valid solutions have not been found, there are a
number of useful limiting approximations which can be derived by first solving the Lapl
transformed equations. Letting $\bar{C}_0$ denote the transformed concentration, and requirin
both C_0 and the radial component of the diffusion current be continous at $r = a$, then in t
dimensions one has that (Carslaw and Jaegar, 1959)

$$\bar{C}_0 = \frac{C'}{p} - \frac{q_2 D_2 K_1 (q_2 a) I_0 (q_1 r)}{p \Delta} \qquad r < a \qquad\qquad 4.18$$

$$\bar{C}_0 = \frac{C'}{p} \frac{q_1 D_1 I_1 (q_1 a) K_0 (q_2 r)}{\Delta} \qquad r \geq a \qquad\qquad 4.19$$

where

$$\Delta \equiv q_1 D_1 I_1 (q_1 a) K_0 (q_2 a) + q_2 D_2 K_1 (q_2 a) I_0 (q_1 a) \qquad\qquad 4.20$$

$$q_1 \equiv \left(\frac{p}{D_1}\right)^{\frac{1}{2}} \qquad q_2 \equiv \left(\frac{p}{D_2}\right)^{\frac{1}{2}}$$

where C' is the total concentration of ith species, p the transform variable and I_n and K
are modified Bessel and Hankel functions of order n.

If one now defines the parameters

$$X_1 \equiv \left(\frac{a^2}{4D_1 t}\right)^{\frac{1}{2}} \qquad\qquad 4.21$$

$$X_2 \equiv \left(\frac{a^2}{4D_2t}\right)^{\frac{1}{2}}$$

4.22

three time regimes can be distinguished. For short times X_1 and X_2 are both very much greater than one, whereas for long times they are much less than one. In the former limit, using the asymptotic expansions of the Bessel and Hankel functions for large values of the argument one finds that

$$C_0(r,t) = \left(\frac{(C')^2 \, a}{r(1+\alpha)^2}\right)^{\frac{1}{2}} \left\{\left[1 + \frac{(r-a)\,G}{2a}\right] \mathrm{erfc}\left(\frac{r-a}{2\sqrt{D_2t}}\right)\right.$$

4.23

$$\left. - G\sqrt{\frac{D_2t}{\pi a^2}} \; \exp\left[-\frac{(r-a)^2}{4D_2t}\right]\right\}$$

where

$$\alpha \equiv \left(\frac{D_2}{D_1}\right)^{\frac{1}{2}}$$

and

$$G \equiv \frac{\alpha}{1+\alpha}\left[1 + (D_1/D_2)^{\frac{1}{2}}\right] - \frac{r-a}{4r}$$

$\mathrm{erfc}(x)$ being the co-error function (Abromowitz and Segun, 1965).

With $D_1 = 1.7 \times 10^{-7}$ cm^2/sec (Suzuki and Deutsch, 1967), and a = 0.2 cm, equation 4.21 requires that the time of the experiment be less than 16.3 hours. Notice that although an experiment lasting several hours may be considered short in terms of the requirements for the validity of the Bessel function expansions, it is long compared to the relaxation times of the chemical reactions $(k_{-1}^{-1} \lesssim$ sec). The latter condition was used in deriving equations 4.6 and 4.7 while the former was used in solving them. Thus equation 4.23 is expected to be adequate when the time of the experiment, T, is constrained according to

$$1 \; \mathrm{sec} < T < 16.3 \; \mathrm{hours}$$

As an alternative limit, one can consider X_1 and X_2 both less than one. Then (Goldstein et al., 1975)

$$C_0(r,t) = \frac{C' \, a^2}{4D_2t}\left[1 - \frac{a^2}{4D_2t}\left(\frac{r^2}{a^2} + \gamma - \frac{D_2}{2D_1}\right)\right]$$

4.24

where $\gamma = 0.5772$

It is evident that, if $D_2 \ll D_0$, there may be a long intermediate time range in which neither expansion is valid. For this case it has not been possible to obtain a result which becomes asymptotically exact. However, a useful approximation can be found by assuming $aq_1 \sim \epsilon$, $aq_2 \sim \frac{1}{\epsilon}$ and substituting the appropriate asymptotic expansions in equation 4 Then, keeping only the leading term

$$\bar{C}_0 = \frac{C' a^2}{2D_2 q_2} \left(\frac{a}{r}\right)^{\frac{1}{2}} \exp\left[-aq_2 \left(\frac{r}{a} - 1\right)\right] \qquad 4.25$$

Taking the Laplace transforms, one has that (Goldstein, et al., 1975)

$$C_0 (r,t) = \frac{C'}{1 + K_1 \rho_0} \left(\frac{a^3}{4\pi D_2 \, rt}\right)^{\frac{1}{2}} \exp\left[-\frac{(r-a)^2}{4D_2 \, t}\right] \qquad 4.26$$

In Table I, $C_0 (r,t)$ at eighty-eight hours obtained using equation 4.26 is compared with results obtained by numerical inversion of equation 4.19. The well radius was taken as 0.21 cm, D_1 as 1.7×10^{-7} cm^2/sec, D_0 as 4.6×10^{-7} cm^2/sec and D_2 as 3.1×10^{-9} cm^2/sec.

TABLE I

r (cm)	$C_0(r,t)/C'$	
	Exact	Approximate
0.21	0.119	0.127
0.23	0.096	0.110
0.25	0.062	0.078
0.27	0.032	0.045
0.29	0.014	0.021
0.31	0.005	0.008

Finally, it is possible to express the general solution for arbitrary time as an integral over the real axis. The result is

$$C_0 = \frac{2C' D_1 \sqrt{D_2}}{\pi} \int_0^\infty \frac{\exp(-D_1 u^2 t) \, J_1 (ua) \, [J_0 (\xi ur) \, \phi(u) - Y_0 (\xi ur) \, \psi(u)] \, du}{u[\phi^2 (u) + \psi^2 (u)]} \qquad 4.27$$

where

$$\psi(u) = D_1\sqrt{D_2}\ J_1\ (au)\ J_0\ (\xi\ au) - D_2\sqrt{D_1}\ J_0\ (au)\ J_1\ (\xi au)$$

$$\phi(u) = D_1\sqrt{D_2}\ J_1\ (au)\ Y_0\ (\xi au) - D_2\sqrt{D_1}\ J_0\ (au)\ Y_1\ (\xi au)$$

and
$$\xi \equiv \frac{D_1}{D_2}$$

As I mentioned earlier, although equations 4.6 and 4.7 were derived under local equilibrium conditions, their validity is actually only constrained by the more general condition that the time of the experiment be longer than the reciprocal of the reverse rate constant. For example, in a twenty-hour experiment, even a reverse rate constant is slow as $10^{-4}\ sec^{-1}$ satisfies this constraint, although such slow dissociations clearly would not satisfy local equilibrium conditions. The physical basis of this is clear. As long as there are many associations and dissociations within the time of the experiment, the chemical process will be describable by some effective equilibrium constant K'. One would thus obtain $D_2 = D_0/\ (1 + K'\ \rho_0)$ where K' is a function of the kinetic parameters of the elementary steps, and will reduce to K_1 only under special conditions (see Chapter 5).

When local equilibrium does in fact prevail, the equations may simplify further since rapid dissociations might imply small enough values of K_1 so that $K_1\ \rho_0 \lesssim 1$ and consequently $D_2 \cong D_0$. With $\rho_0 = 2 \times 10^{12}$ sites/cm^3 this condition on D_2 will be approximately valid for $K_1 \lesssim 2.5 \times 10^{-13}\ \dfrac{cm^3}{molecule} = 1.5 \times 10^8\ M^{-1}$. For univalent associations it is not unusual to have mean affinities less than $10^8\ M^{-1}$ (Werblin and Siskind, 1972) so it is reasonable to expect that there will often be populations for which $D_2 \cong D_0$. With this condition, equation 4.17 simplifies somewhat. In particular

$$N(r,t) = 2E \sum_i C_{0i}\ K_{1i} = 2E \sum_i C_i'\left(\frac{C_{0i}}{C_i'}\right)K_{1i} = 2EC_T \sum_i \frac{C_i'}{C_T}\frac{C_{0i}}{C_i'}\ K_{1i} \qquad 4.28$$

where the sum is over all antibody affinities, C_{0i} being the concentration of diffusing antibodies in the ith subpopulation having affinity K_{1i} and

$$C_T = \sum_i C_i'$$

With $D_2 = D_0$ for the entire population, it is evident that $\frac{C_{0i}}{C'_i}$ is affinity-independent and can therefore be removed from the sum. In addition, since the average value of K is defined as

$$\bar{K} = \sum_i \frac{C'_i}{C_T} K_{1i}$$

equation 4.28 can be written as

$$N(r,t) = 2\bar{K} \ E \ C_T \left(C_0/C' \right) \qquad 4.29$$

where $\frac{C_0}{C'}$ is given by equation 4.27. By the same procedure, the intermediate time resu equation 4.26 becomes

$$N(r,t) = 2E \sum_i C_{0i} K_{1i} = 2E \sum_i C'_i \left(\frac{C_{0i}}{C'_i} \right) K_{1i} = 2E \ C_T \sum \frac{C'_i \ C_{0i} \ K_{1i}}{C_T \ C'_i}$$

or

$$N(r,t) = 2\bar{K} \ E \ C_T \left(\frac{a^3}{4\pi D_0 \ rt} \right)^{\frac{1}{2}} \exp \left\{ - \frac{(r-a)^2}{4 \ D_0 \ t} \right\} \qquad 4.30$$

4.4 Applications

4.4.1 Estimation of the Diffusion Coefficient

It is evident that the plaque radius depends not only on time and antiserum con tration, but also on the antibody diffusion coefficient and affinity distribution. The effe of the latter are clearly illustrated by the experiments of Hiramoto and his colleagues (1971), who found that plaque radius decreases as RBC concentration increases. This i agreement with equation 4.8, according to which the diffusion coefficient is reduced by factor $(1 + 2K_1 \rho_0)$. This results in a decreased concentration of diffusing antibody, a hence a reduction in the concentration of bound antibody and in the probability of lysis The effect is qualitatively simple and suggests a sensitivity of the results to changes in fusion coefficient.

Plaque radii have also been measured as a function of initial antiserum concen and the analysis of these experiments is especially fruitful. If fC_T parameterizes the a serum concentrations in different wells ($0 < f \leq 1$) at $t = 0$, then r_p (f) is obtained experimentally. The form of this function can be readily predicted. For example, from equation 4.30, one finds that

$$\tfrac{1}{2} \ln \left[\frac{(f^2\, a\,)}{r_p t} \right] = \left(\frac{a^2}{4D_0 t} \right)^{\frac{1}{2}} \left(\frac{r_p}{a} - 1 \right)^2 + \ln\Omega \qquad \qquad 4.31$$

where

$$\Omega \equiv \left(\frac{4\pi\, D_0}{a^2} \right)^{\frac{1}{2}} \frac{N^*}{\overline{KE}\, C_T} \qquad \qquad 4.32$$

Consequently, a plot of the left side of equation 4.31 against $\frac{1}{t} \left(\frac{r_p}{a} - 1 \right)^2$ should be a straight line with slope $\frac{a^2}{4D_0}$.

There are, unfortunately, no experiments currently available which meet the conditions of the theory. The closest one can come is the data published by Hiramoto and his colleagues using unhaptenated RBCs. In one series of experiments, four sets of measurements were performed with

C_T = (96.20, 93.31, 69.26 and 28.67) µg/ml.

For each C_T, eight serial dilutions were made with $\frac{1}{8} \leq f \leq \frac{1}{1024}$. In Figure 4.2 all the experimental results have been plotted according to equation 4.31 by taking f = 1 at C_T = 96.20 and correspondingly smaller values of f at other values of C_T (Goldstein and DeLisi, 1975). Thus at C_T = 93.31, $f = \frac{93.31}{96.20}$, etc. This gives a total of 32 data points with 2.91 x $10^{-4} \leq f \leq 1$. Also shown on the same plot is a theoretical curve calculated numerically and chosen to fit the data at large radii. The best fit was obtained for $D_2 = 4 \times 10^{-8}$ cm^2/sec. For comparison the diffusion coefficient of IgM in water is 1.7×10^{-7} cm^2/sec.

It is apparent that, although the points appear to lie on a straight line at large r_p, deviation of theory from experiment becomes dramatic as plaque size decreases. This divergence may be understood in terms of diffusion coefficient heterogeneity. Since antibodies with $K_1 \rho_0 \lesssim 1$ have $D_2 \sim D_0$, whereas those with $K_1 \rho_0 > 1$ have $D_2 < D_0$, the rate of antibody diffusion is affinity-dependent. Consequently, one would expect that the radii

of large plaques will be determined by relatively low affinity antibodies, since these dif
fuse most readily. This would tend to be true particularly at high antiserum concentrati

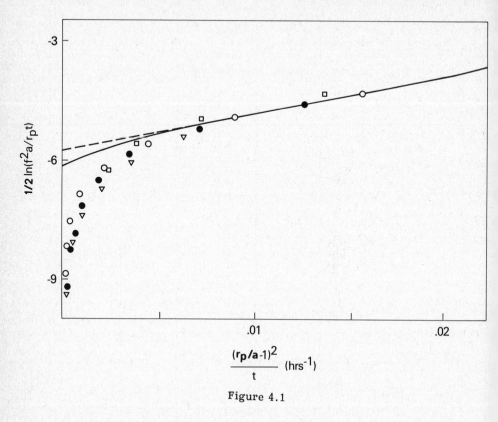

$$\frac{(r_p/a\text{-}1)^2}{t} \ (hrs^{-1})$$

Figure 4.1

As the concentration is made more dilute, plaques will, of course, become smal
since there will be less antibody reaching a given distance in a fixed amount of time. T
low affinity antibodies will still diffuse the furthest and their concentration will be redu
by the same fraction as the antibodies in every other subpopulation. However, since th
were in small quantity to begin with (i.e., at the extreme, low amplitude end of the dist
bution), the concentration now may be so low that there are not enough of them to effect ly
In this case the edge of the plaque would be determined by a slightly higher affinity sub
population. As f is reduced further, the plaque radius decreases and is determined by
still higher affinity subpopulation, and so on. Thus the value of D_2, which is reflected
the slope of the plot, changes as the subpopulation which determines the plaque radius ch
ges. As $K_1 \rho_0$ continues to increase to values larger than one, the change should show as
initial slow shift in the slope which gradually increases as f decreases, precisely the be

havior shown in Figure 4.1.

4.4.2 Estimation of Average Affinity

According to equations 4.31 and 4.32, the point at which the asymptote in Figure 4.1 (dashed line) intercepts the Y axis is related to $\bar{K}_1$. Therefore, if N^* is known, $\bar{K}_1$ can be determined. The utility of such a procedure resides in the fact that N^* is, in effect, a universal constant; i.e., once it is determined for any particular experiment, it can be used in the analysis of any plaque assay. The evaluation of $\bar{K}_1$ by this method is, however, limited to low affinity antibody populations for which equations 4.6 and 4.7 hold with $D_2 \overset{\sim}{=} D_0$. Under these conditions, provided the time of the experiment is chosen appropriately, the numerical solution to equation 4.19 does not differ significantly from the approximate solution given by equation 4.30.

5

THE THEORY OF PLAQUE GROWTH KINETICS FOR A
LYMPHOCYTE SOURCE

5.1   Introduction

Antibody heterogeneity is inextricably involved in the specificity and dynamics of the humoral response, and hence the latter cannot be fully understood without detailed knowledge of the former. At the same time, however, the presence of heterogeneity complicates the analysis of conventional experiments which would be used to elucidate its features. The problem might, in principle, be avoided by studies at a single cell level. That means using a method for isolating, enumerating, and identifying individual antibody secreting cells and adapting it to physical chemical studies.

Identification of single antibody secreting cells (ASCs) can, in fact, be accomplished by a hemolytic plaque experiment using live lymphocytes from an immunized donor (Jerne et al., 1963; Ingraham, 1963). The method is similar to the plaque technique described in Chapter 4, but with lymphocytes rather than an antiserum well as the source of antibodies. Thus on the day of the test one obtains spleen cells from an immunized donor and mixes them in an agar gel with appropriately haptenated RBCs. During incubation, which usually lasts about 50 minutes, a small fraction of lymphocytes synthesize antibody and secrete it into the surrounding medium where it interacts specifically with the haptenated RBCs. Flooding the plate with complement results in clear, well-separated plaques, each centered on an ASC. In this procedure, each plaque is produced by a homogeneous antibody population (Chapter 1), and this makes the analysis of the technique simpler and more profitable than that the well method.

There are several types of information one would like to obtain from these experiments. The most obvious is a characteristic of the intensity of a response, i.e., the number of antibody secreting cells. This can be studied as a function of immune parameters: time after immunization, antigen dose, degree of response, etc. Mathematical modeling is, of course, unnecessary for this sort of study and immunologists have, in fact, been doing experiments of this type for a number of years. There is, however, additional information whose relationship to observables is not so apparent.

It is evident that the size of a plaque will depend upon a number of factors, including incubation time, secretion rate, medium viscosity, antibody-epitope reaction rate constants, and epitope density. Except for secretion rate and reaction rate constants, the other parameters are known or controllable. So for any particular plaque, there are three unknowns: the forward and reverse rate constants, and the secretion rate. The forward rate constant is, to a good approximation, invariant and is therefore not expected to contribute significantly to differences among plaques. The reverse rate constant and secretion rate, however, should vary considerably from cell to cell. The former will reflect the diversity of antibody combining sites while the latter will reflect differences in state of differentiation and perhaps the cell cycle. By developing a mathematical model, one would hope to be able to delineate conditions under which variations in plaque characteristics will preferentially reflect only a single parameter distribution. If this could be done, it would provide a relatively simple method for obtaining the information which is required to construct physical chemical theories of the immune response.

5.2   A Diffusion-Reaction Model

The ASC is considered to be a point source embedded in an infinite, isotropic n-dimensional ($n = 1$, 2, 3) medium. As antibody diffuses from the lymphocyte, it may encounter RBC epitope and bind univalently, and this reaction can be followed by dissociation or formation of a second bond. Consequently, three reactive species must be considered: diffusing antibodies at concentration C_0, singly bound antibodies at concentration C_1, and doubly bound antibodies at concentration C_2. Hence (DeLisi, 1975c)

$$\frac{\partial C_0}{\partial t} = D_0 \nabla^2 C_0 - \frac{\partial}{\partial t}(C_1 + C_2) \tag{5.1}$$

$$\frac{\partial C_1}{\partial t} = 2k_1 C_0 p_0 - k_{-1} C_1 - k_2' C_1 + 2k_{-2} C_2 \tag{5.2}$$

$$\frac{\partial C_2}{\partial t} = k_2' C_1 - 2k_{-2} C_2 \tag{5.3}$$

Equations 5.1 - 5.3 are to be solved subject to the initial conditions

$$C_0(r,0) = 0 \hspace{6cm} 5.4$$

$$C_1(r,0) = 0$$

$$C_2(r,0) = 0$$

and the boundary condition that C_0 vanish isotropically at a large distance from the sourc

$$\lim_{r \to \infty} C_0(r,t) = 0 \hspace{5cm} 5.5$$

In addition, there is a boundary condition at the origin, viz., that the diffusion current equals the rate of antibody production there. For example, in three dimensions this is

$$\lim_{r \to 0} 4\pi r^2 D_0 \nabla C_0 = S(t) \hspace{4cm} 5.6$$

where $S(t)$ is the number of antibodies secreted per second at time t by the cell under co
sideration. Equation 5.1 describes the time rate of change of free antibodies as the resul
of diffusion (first term on the right) and chemical reaction (second term on the right). E
tions 5.2 and 5.3 are the chemical reaction equations for singly and doubly bound antibo
The first and last terms on the right in equation 5.2 represent the formation of C_1 by, re
spectively, association of diffusing antibody with RBC epitope, and dissociation of doubl
bound antibody. The two central terms represent its disappearance by dissociation from
the RBC and conversion to C_2, respectively. The first term on the right in equation 5.3
represents conversion of C_1 to C_2 and the second conversion of C_2 to C_1. The equations
implicitly treat bivalent antibody, but for some of the special cases to be considered, the
effect of increasing valence beyond two will simply be to change the constant multiplying
k_1.

5.3   Solutions Under Limiting Conditions

5.3.1   The n-Dimensional Green's Function

Rather than attempting to solve these equations, it is worthwhile to first conside
the time scales involved in the various reaction and diffusion processes, and their relati
to the time of the entire experiment. It often happens that, depending upon how the expe

65

mental system is prepared, biases are produced against certain types of reactions and in favor of others due to the dominance of a certain kinetic parameter or combination of parameters. For the system under consideration, k'_2 is of particular interest since it determines the relative importance of different kinetic pathways and, equally significantly, is experimentally controllable, reflecting epitope density and hinge flexibility (antibody class).

One limit of interest is suggested by the results of the phage neutralization analysis (Chapter 2), from which it was concluded that: (1) k'_2 is very large compared to either k_{-1} or k_{-2} and (2) the effective reverse rate constant for a multivalently bound antibody is about 10^{-4} sec^{-1}. These results imply that under appropriate conditions, the reaction is irreversible and rate limited by the bimolecular step.

As an alternative extreme, one may consider, as in Chapter 4, sparsely haptenated RBCs, so that multisite attachment is difficult. However, even if the mean spacing between epitopes exceeds the maximum distance between the combining sites of an antibody, intramolecular binding may still be possible since the fluid membrane (Singer and Nicholson, 1972) permits epitopes to diffuse close to one another. Consequently, if a singly bound antibody remains attached for a sufficiently long time, an epitope will diffuse close enough so that the free combining site can interact with it, thus establishing a second bond.

It is clear that the relevant quantity to consider in evaluating the influence of this effect is the ratio of the mean dissociation time for singly bound antibody, to the mean time for establishing a second bond. Dissociation of singly bound antibody occurs rapidly, typical times being (0.01 - 1) sec. The rate for establishing a second bond when epitope diffusion is required is not known experimentally, but has been estimated to be $\gtrsim 10$ sec when there are 10^5 epitopes/cell (Bell, 1974). Consequently, one would expect many associations and dissociations before a successful cross linking reaction (assuming the mean distance between epitopes cannot be spanned by the antibody). In any event, for sufficiently sparse haptenation, cross linking is expected to be significantly slower than dissociation and a local equilibrium assumption can be pursued.

The above limits to the reaction equations are somewhat analogous to the pseudo steady state (Briggs and Haldane, 1925) and pseudo equilibrium (Michaelis and Menton, 1913) hypotheses used to describe the kinetics of numerous biochemical processes (Heineken

et al., 1967). In both cases the time rate of change of C_1 due to chemical reaction is zer
in the former limit because of local steady state conditions; in the latter because of local
equilibrium. Thus, with irreversible bivalent binding

$$C_1 \approx \frac{2k_1 \, C_0 \, \rho_0}{k_{-1} + k_2'} \qquad \qquad 5.7$$

and

$$\frac{\partial C_1}{\partial t} \approx \frac{2k_1 \rho_0}{k_{-1} + k_2'} \frac{\partial C_0}{\partial t} \qquad \qquad 5.8$$

Substituting equation 5.7 into equation 5.3

$$\frac{\partial C_2}{\partial t} = \frac{2k_1 \, k_2' \, \rho_0 \, C_0}{k_{-1} + k_2'} \qquad \qquad 5.9$$

Consequently, equation 5.1 becomes

$$\frac{1}{D} \frac{\partial C_0}{\partial t} = \nabla^2 C_0 - \frac{C_0}{L^2} \qquad \qquad 5.10$$

where

$$D \equiv \frac{D_0}{1 + 2k_1 \rho_0 / (k_{-1} + k_2')} \qquad \qquad 5.11$$

and

$$L^2 \equiv \frac{D_0 \, (k_{-1} + k_2')}{2 \, k_1 \, k_2' \, \rho_0} \qquad \qquad 5.12$$

The Green's function for equation 5.10

$$G(t,r;t') = \frac{(D/D_0) \, S(t')}{[4\pi D \, (t - t')]^{n/2} \Delta(n)} \exp\left\{ - \frac{r^2}{4D \, (t - t')} - \frac{D \, (t - t')}{L^2} \right\} \qquad 5.13$$

where n (= 1, 2, 3) labels the dimensionality and Δ (n) is a geometrical factor equal to the cross sectional area of the capillary for n = 1, the thickness of RBC layer for n = 2, and one for n = 3. G may be interpreted as the concentration of free antibodies at time t and distance r, produced by a cell which begins to secrete at t' and thereafter gradually becomes depleted. Equation 5.13 is therefore expected to apply to experiments in which transcription or translation has been blocked by a metabolic inhibitor. In fact, under such conditions the number of antibodies bound per RBC is found by integrating equations 5.8 and 5.9 (with C replaced by G), adding the results, and dividing the sum by the concentration of RBCs. Thus, in three dimensions

$$
N = \frac{2K_1 \, EG}{1 + k_2'/k_{-1}} + \frac{2k_2' \, K_1 \, E \, (D/D_0) \, S(t')}{1 + k_2'/k_{-1}} \int_{t'}^{t} \frac{\exp\left\{ \frac{r^2}{4D(t-\tau)} - \frac{D(t-\tau)}{L^2} \right\}}{[4\pi D(t-\tau)]^{3/2}} d\tau \qquad 5.14
$$

or

$$
N = \frac{2K_1 \, EG}{1 + k_2'/k_{-1}} + \frac{k_2' \, K_1 \, E \, S(t')}{(1 + k_2'/k_{-1}) \, 4\pi \, D_0 r} \left\{ e^{-r/L} \left[\mathrm{erf}\left(\frac{\sqrt{Dt}}{L} - \frac{r}{2\sqrt{Dt}} \right) \right. \right.
$$

$$
\left. - \mathrm{erf}\left(\frac{\sqrt{Dt'}}{L} - \frac{r}{2\sqrt{Dt'}} \right) \right] + e^{r/L} \left[\mathrm{erf}\left(\frac{\sqrt{Dt'}}{L} + \frac{r}{2\sqrt{Dt'}} \right) \right.
$$

$$
\left. \left. - \mathrm{erf}\left(\frac{\sqrt{Dt}}{L} + \frac{r}{2\sqrt{Dt}} \right) \right] \right\} \qquad 5.15
$$

erf(x) being the error function.

In order to proceed further, the source distribution function, S(t), must be specified. The simplest assumption is that it is constant throughout the entire 50-minute incubation period. Such constancy is in fact not difficult to achieve experimentally and is expected when optimal conditions are closely adhered to.

5.3.2   The Solution in Three Dimensions

Since there is no general n-dimensional solution for anything other than an impulsive time source, each dimension will be considered separately. With n = 3, and the cell secreting constantly from t' = 0 to t' = t, integration of equation 5.13 gives

$$C_0 = \frac{S_0 \exp(-r/L)}{8\pi\, D_0 r}\left[1 + \mathrm{erf}\!\left(\frac{\sqrt{Dt}}{L} - \frac{r}{2\sqrt{Dt}}\right)\right]$$
$$+ \frac{S_0 \exp(r/L)}{8\pi D_0 r}\left[1 - \mathrm{erf}\!\left(\frac{\sqrt{Dt}}{L} + \frac{r}{2\sqrt{Dt}}\right)\right] \tag{5.16}$$

where S_0 is the number of antibodies secreted per second.

The number of antibodies bound per RBC is

$$N = \frac{2K_1 E C_0}{1 + k_2'/k_{-1}} + \frac{2k_2' K_1 E}{1 + k_2'/k_{-1}} \int_0^t C_0 \, dt \tag{5.17}$$

where C_0 is given by equation 5.16. Limiting conditions will now be considered more s
fically.

Local Steady State: $\dfrac{k_{-1}}{k_2'} \to 0$

In this case, as soon as the first bond is established, the second one forms instantaneou
Chemical reaction is therefore assumed to be in a steady state and equation 5.17 become

$$N = 2k_1 E \int_0^t C_0\,(r,t')\,dt' \tag{5.18}$$

where $C_0(r,t)$ is given by equation 5.16 with

$$D = D_0 \tag{5.19}$$

and

$$L^2 = \frac{D_0}{2k_1 \rho_0} \tag{5.20}$$

The first term in equation 5.17 has been neglected since $k_{-1} \gg k_1 E$ and, therefore, if

$\dfrac{k_{-1}}{k_2'} \to 0$, so does $k_1 E/k_2'$. Under these conditions the plaque radius, r_p, is obtained b

solving

$$N^* = 2k_1 E \int_0^t C_0 (r_p, t') \, dt' \qquad 5.21$$

Local Equilibrium: $\dfrac{k_2'}{k_{-1}} \to 0$

In this limit only single site attachment is possible and since the time constant for the chemical reaction is very rapid compared to the time required for significant concentration change due to diffusion, local equilibrium is assumed. Therefore, from equations 5.16 and 5.7

$$N = \frac{K_1 E S_0}{2\pi D_0 r} \, \text{erfc} \left(\frac{r}{2 \sqrt{Dt}} \right) \qquad 5.22$$

with

$$D = \frac{D_0}{1 + 2K_1 \rho_0} \qquad 5.23$$

and the plaque radious is given by

$$r_p = \frac{K_1 E S_0}{2\pi D_0 N^*} \, \text{erfc} \left(\frac{r_p}{2 \sqrt{Dt}} \right) \qquad 5.24$$

When $2K_1 \rho_0 \ll 1$, equation 5.22 reduces to the expression used by Jerne et al. (1974).

5.3.3   The Solution in Two Dimensions

I have been unable to obtain a two-dimensional solution corresponding to equation 5.16. However, series solutions can easily be developed. For example, at short times substituting

$$\tau = \frac{r^2}{4D(t - t')} \qquad 5.25$$

into equation 5.13 and integrating, one has that

$$C_0 = \frac{S_0}{4\pi D_0 \Delta (2)} \int_{\frac{r^2}{4Dt}}^{\infty} \frac{\exp \left[-\tau - \left(\frac{r}{2L} \right)^2 \frac{1}{\tau} \right] d\tau}{\tau} \qquad 5.26$$

Expansion of exp $[-\left(\frac{r}{2L}\right)^2 \frac{1}{\tau}]$ leads to a solution in terms of a series in incomplete gamma functions.

$$C_0 = \frac{S_0}{4\pi D_0 \Delta(2)} \sum_{j=0}^{\infty} \Gamma\left(-j, \frac{r^2}{4Dt}\right) \frac{(-1)^j}{j!} \left(\frac{r}{2L}\right)^{2j} \qquad 5.27$$

Under <u>local</u> <u>steady</u> <u>state</u> conditions, equation 5.26 retains its form with D and L given by equations 5.19 and 5.20. In the <u>local equilibrium</u> limit, however, equation 5.27 simplfies since $\frac{r}{2L} \to 0$. Therefore, only the first term in the expansion remains.

$$C_0 = \frac{S_0}{4\pi D_0 \Delta(2)} \Gamma\left(0, \frac{r^2}{4Dt}\right) = -\frac{S_0}{4\pi D_0 \Delta(2)} Ei\left(-\frac{r^2}{4Dt}\right) \qquad 5.28$$

and

$$N = -\frac{K_1 E S_0}{2\pi D_0 \Delta(2)} Ei\left(-\frac{r^2}{4Dt}\right) \qquad 5.29$$

where $-Ei(-x)$ is the exponential integral. The plaque radius is found by solving

$$\frac{2\pi D_0 \Delta(2) N^*}{K_1 E S_0} = -Ei\left(-\frac{r_p^2}{4Dt}\right) \qquad 5.30$$

Equation 5.27 should be contrasted to the result for the <u>global steady</u> state, <u>i.e.</u> the $t = \infty$ limit. In that case equation 5.26 can be written as a zero order modified Bessel function of the second kind and one finds that

$$N = \frac{S_0 K_1 E K_0(r/L)}{2\pi D_0 \Delta(2)} \qquad 5.31$$

$K_0(r/L)$ being the Bessel function.

5.4 Models Allowing Reversible Reactions

Although the derivations of equations 5.22 and 5.29 assume local equilibrium, th same equations hold even if chemical reaction is slow, so long as $T >> (2k_{-2})^{-1}$, where is the time of the experiment. The reason is that under this condition equations 5.2 and 5.3 can be replaced by a single chemical reaction equation that always has a fast effectiv reverse rate constant. This is easily demonstrated by using operational calculus. Let $C_B = C_1 + C_2$. Then the problem is to find forward and reverse rate constants, k_f and k_r, such that equations 5.2 and 5.3 can be replaced by

$$\frac{\partial}{\partial t} C_B = k_f C_0 P_0 - k_r C_B \qquad 5.32$$

Adding equations 5.2 and 5.3 and introducing the operator $p = \frac{\partial}{\partial t}$, one has that

$$p(C_1 + C_2) = 2k_1 C_0 P_0 - k_{-1} C_1 \qquad 5.33$$

Also from equation 5.3

$$(p + 2k_{-2}) C_2 = k_2' C_1 \qquad 5.34$$

Adding $(p + 2k_{-2}) C_1$ to both sides of equation 5.34

$$C_B (p + 2k_{-2}) = C_1 (p + 2k_{-2} + k_2') \qquad 5.35$$

or

$$C_B = C_1 \left[1 + \frac{K_2'}{1 + p/2k_{-2}} \right] \qquad 5.36$$

where

$$K_2' \equiv \frac{k_2}{2k_{-2}}$$

Evidently the time scale is determined by $2k_{-2}t$. Assuming this is sufficiently large to neglect powers of $p/2k_{-2}$ higher than the first, equation 5.36 becomes

$$C_1 = \frac{C_B}{1 + K_2'} \left[1 + \frac{K_2'}{2(1 + K_2') k_{-2}} p \right] \qquad 5.37$$

Substituting equation 5.37 into equation 5.33

$$\frac{\partial C_B}{\partial t} = 2k_1 C_0 P_0 - \frac{k_{-1}}{1 + K_2'} \left[1 + \frac{K_2'}{2(1 + K_2')k_{-2}} \frac{\partial}{\partial t} \right] C_B \qquad 5.38$$

Combining terms in $\frac{\partial C_B}{\partial t}$, one has equation 5.32 with

$$k_f \equiv \frac{2k_1 (1 + K_2')^2}{(1 + K_2')^2 + k_{-1} K_2'/2k_{-2}} \qquad 5.39$$

$$k_r = \frac{k_{-1}(1 + K_2')}{[(1 + K_2')^2 + k_{-1}K_2'/2k_{-2}]}$$

5.40

and

$$K_{eff} \equiv \frac{k_f}{k_r} = 2K_1(1 + K_2')$$

Thus equations 5.2 and 5.3 can be replaced by a single differential equation with k_f and given by equations 5.39 and 5.40. Equations 5.1 and 5.32 are formally the same as the ec tions originally written by DeLisi and Bell (1974) to describe plaque growth.

To understand why local equilibrium is expected to be a good approximation, noti that $k_r \leq k_{-1}$ and therefore irrespective of how large K_2' is, so long as $2k_{-2} t \lesssim 1$, the effe tive reverse rate constant is rapid (i.e., greater than k_{-1}). Hence, either equation 5.2 or 5.29 should be reasonably applicable with D given by equation 5.23 but with K_{eff} re- placing $2K_1$.

Up to this point reverse reactions have not been treated explicitly but have eithe been disregarded (irreversible binding) or have appeared only through the equilibrium co stant. However, when plaque growth can be modeled by equations 5.1 and 5.32, a solut for the Green's function can be obtained, either by Laplace transforming the diffusion eque or by a more fundamental analysis of the diffusion process. In the latter method (Jerne, 1974), the probability, P, that an antibody released at time zero will be attached to an R at (r,t) is written as

$$P = \int_0^t P_\tau \, P_r \, d\tau \, d\nu$$

5.41

where $P_\tau d\tau$ is the probability that an antibody bound at t,r will have diffused freely dur a time $\tau \leq t$, and $P_r d\nu$ is the probability that a molecule diffusing freely for time τ is pre at r.

P_τ can be calculated by dividing τ into $\tau/\Delta\tau$ intervals, where multiple association and dissociations may be neglected during the very small time period $\Delta\tau$. If there are j + associations and j dissociations during τ, there will be many ways in which these may be distributed among the total number of time intervals. In particular, W_f, the number of we in which $j + 1$ periods of free diffusion can occur, is

$$W_f = \frac{(\tau/\Delta\tau - 1)!}{j! \, (\tau/\Delta\tau - 1 - j)!} \cong \frac{(\tau/\Delta\tau)^j}{j!} \qquad 5.42$$

Similarly the number of ways in which the j bound periods can occur is

$$W_b = \frac{[\,(t - \tau)/\Delta\tau\,]^j}{j!} \qquad 5.43$$

Thus the total number of ways of undergoing $j + 1$ periods of free diffusion and j periods of attachment is $W_f W_b$, and for any __particular__ way, the probability that the molecule will be bound at time t is

$$(k_1 P_0 \Delta\tau)^{\,j + 1} \, (k_{-1} \Delta\tau)^j \, (1 - k_1 P_0 \Delta\tau)^{\tau/\Delta\tau} (1 - k_{-1}\Delta\tau)^{\,(t - \tau)/\Delta\tau} \qquad 5.44$$

The first two terms are the probabilities of $j + 1$ attachments and j detachments, respectively; the third, the probability of not attaching during $\tau/\Delta\tau$ intervals, and the fourth, the probability of not detaching during $(t - \tau)/\Delta\tau$ intervals. Multiplying this product by $W_f W_b$, summing over j, and letting $\Delta\tau$ become infinitesimal, one has that

$$P_\tau = k_1 P_0 \exp\left[- k_1 P_0 \tau - k_{-1} \, (t - \tau)\right] \sum_{j = 0}^{\infty} \left(\frac{x^j}{j!}\right)^2 \qquad 5.45$$

where $x^2 = k_1 P_0 k_{-1} \tau \, (t - \tau)$. The sum will be recognized as a representation of $I_0 \, (2x)$, the zero order modified Bessel function of the first kind.

By definition, $P_r \, dv$ is the probability that a molecule diffusing __freely__ during time τ will be present at r. Hence, in n dimensions

$$P_r \, dv = \frac{1}{(4\pi \, D_0 \tau)^{n/2}} \, \exp\left(- \frac{r^2}{4D_0\tau}\right) dv \qquad 5.46$$

Consequently

$$P = \int_0^t \frac{k_1 P_0 \, dv}{(4\pi D_0\tau)^{n/2}} \, \exp\left\{- \frac{r^2}{4D_0\tau} - k_1 P_0 \tau - k_{-1} \, (t - \tau)\right\} I_0 \, (2x) \, d\tau \qquad 5.47$$

The Green's function for the system of equations 5.1 and 5.32, i.e., for the numb

of antibodies bound per RBC is P S_0 /(number of RBCs in dv) or

$$N = k_1 E S_0 \int_0^t \frac{\exp\left[-\frac{r^2}{4D_0\tau} - \frac{D_0\tau}{L^2} - k_{-1}(t - \tau) \right] I_0(2x)}{(4\pi D_0\tau)^{n/2} \Delta(n)} \, d\tau \qquad 5.48$$

It is straightforward to extend this analysis to models with more than one chemical reaction

equation.

Although these equations were developed explicitly for bivalent antibodies, the

effect of valence under limiting conditions is easy to assess. In particular, it enters the

forward rate constant as a multiplicative factor so that if an f valent (f > 2) rather than a

bivalent antibody were considered, the main change would be a factor f rather than 2 mul

tiplying k_1 and K_1.

5.5   A Diffusion-Transport-Reaction Model.   Electrophoresis

For many experimental systems, pH conditions may be set so that antigen is charge

Migration into an antiserum bed may then be facilitated by applying an electric field paralle

to the face of the petri dish. Alternatively, if antigen is immobilized, as in the Jerne plaque

technique, conditions may be set so that antibody is transported by the field. In the latter

case pH must be restricted to about 7.4 ± 0.2 if the lymphocytes are to remain healthy. How

ever, since the isoelectric point of most antibodies is about one pH unit higher, a large num

of plaques may be affected by the field. Movement of this sort — of charged particles and

macromolecules under the influence of an electric field — is known as electrophoresis.

Modification of the theory to take account of the presence of crossed electric field

may be effected by adding transport terms to equation 5.1. Thus in terms of the cartesian

coordinates x_1, x_2 and x_3,

$$\frac{\partial C_0}{\partial t} = D_0 \nabla^2 C_0 - \sum_{i=1}^{n} v_i \frac{\partial C_0}{\partial x_i} - \frac{\partial}{\partial t}(C_1 + C_2) \qquad n = 1,2,3 \qquad 5.49$$

where

$$v_i = \mu_i E_i \qquad 5.50$$

μ_i and E_i being the electrophoretic mobility and field intensity in the ith direction. Equation 5.49 clearly also applies to a single, arbitrarily directed field. Since the lymphocyte is considered a point source, its geometry plays no role.

The most direct way to proceed is by introducing the Galilean transformation

$$x_i' = x_i - v_i t \qquad\qquad i = 1,2,3 \qquad\qquad 5.51$$

$$t' = t$$

Then in the primed system, equation 5.49 becomes

$$\frac{\partial C_0}{\partial t} = D_0 \nabla^2 C_0 - \frac{\partial}{\partial t}(C_1 + C_2) \qquad\qquad 5.52$$

where the Laplacian is understood to be taken with respect to the primed coordinates. Equation 5.52 has the same form as equation 5.1. Hence, for a lymphocyte which begins secreting antibody at $t = 0$, the n-dimensional Green's function in the unprimed system is

$$G(x_1, x_2, x_2; t) = \frac{(D/D_0)S_0}{(4\pi Dt)^{n/2}\Delta(n)} \exp\left\{ -\frac{\sum\limits_{i=1}^{n}(x_i - v_i t)^2}{4Dt} - \frac{Dt}{L^2}\right\} \qquad 5.53$$

$$= \frac{(D/D_0)S_0 \exp\left(\frac{\sum\limits_{i=1}^{n} x_i v_i}{2D}\right)}{(4\pi Dt)^{n/2}\Delta(n)} \exp\left\{ -\frac{r^2}{4Dt} - \frac{Dt}{\gamma^2}\right\}$$

where

$$\frac{1}{\gamma^2} \equiv \frac{1}{L^2} + \left(\frac{v}{2D}\right)^2 \qquad\qquad 5.54$$

Equation 5.53 will also be the solution to the problem if proper care is not taken in setting pH conditions. Actually the use of dead lymphocytes may not be so sterile a procedure as one might think since there is no reason to believe that intracellular antibody concentration is any less an indicator of the state of differentiation than secretion rate. Since the equations are considerably simpler in this case, the use of dead lymphocytes along with an electric field to facilitate antibody transport from their interior may be fruitful.

The concentration of free antibodies at time t for a continuous source is

$$
C_0 = \frac{(D/D_0)\, S_0 \exp\left(\dfrac{\sum\limits_{i=1}^{n} x_i v_i}{2D}\right)}{\Delta(n)} \int\limits_0^t \frac{\exp\left(-\dfrac{r^2}{4D\tau} - \dfrac{D\tau}{\gamma^2}\right)}{(4\pi D\tau)^{n/2}}\, d\tau \qquad 5.55
$$

Since the integrals in equations 5.55 and 5.14 have the same form, the previous analysis applies equally here. For example, in three dimensions

$$
\begin{aligned}
C_0 = \frac{S_0}{8\pi D_0 r}\; \exp&\left(\frac{\sum\limits_{i=1}^{3} x_i v_i}{2D}\right)\left\{ e^{-r/\gamma}\left[1 + \operatorname{erf}\left(\frac{\sqrt{Dt}}{\gamma} - \frac{r}{2\sqrt{Dt}}\right)\right]\right.\\
&\left. + e^{r/\gamma}\left[1 - \operatorname{erf}\left(\frac{\sqrt{Dt}}{\gamma} + \frac{r}{2\sqrt{Dt}}\right)\right]\right\}
\end{aligned}
$$

$$\qquad 5.56$$

Although this equation is formally the same as equation 5.16, there is a significant difference between them since in the local equilibrium limit, the diffusion length γ can never become infinite with a field present. Hence there will always be some absorption and consequently there will always be a global steady state, even with local equilibrium in two dimensions. More specifically, for $n = 2$ and $t = \infty$, the analogue to equation 5.31 is

$$
C_0 = \frac{S_0\, K_0\,(r/\gamma)}{2\pi D_0\, \Delta(2)}\; \exp\left(\sum\limits_{i=1}^{2} \frac{x_i v_i}{2D}\right) \qquad 5.57
$$

It is not my intention to pursue the development of this analysis here. I wish primarily to indicate that relatively little mathematical effort has been expended in this area and there are a number of unsolved problems involving the use of unidirectional as well as crossed electric fields, and precipitation as well as hemolysis. It may, therefore, be an area appropriate for mathematical and experimental collaboration. The areas of interest, along with their current status, are summarized in Table I. Williams and Chase (1971), Vol. III, pp. 234-294, is a good introduction to the experimental aspects.

TABLE I

CATEGORIES OF IMMUNODIFFUSION

Source	Method of Detection	
	With Field	Without Field
	Hemolysis	
Lymphocyte (homogeneous antibodies)	No experiments; some theory (Goldstein and Perelson, 1976)	Experiment and theory both extensive
	Precipitation	
	Experimental difficulties; no obvious theoretical advantages	
	Hemolysis	
Antiserum (heterogeneous antibodies)	Neither experiment nor theory	Some experiment and theory (see Chapter 4)
	Precipitation	
	Some theory (Aladjem et al., 1962); no experiments	Some theory and experiment
	Hemolysis	
	Experiments not practicable	
Antigen	**Precipitation**	
	Extensive experiments; some theory (Aladjem et al., 1962; Cann, 1975)	Extensive experiments, some theory (Aladjem)

5.6   Predictions of the Theory

5.6.1   Constraints on Parameters

It is useful to begin the discussion by distinguishing control parameters from response parameters. The former characterize the assay and their values are set by the experimentalist (e.g., ρ_0, E and k_2'). The latter characterize the immune response and are to be obtained from an analysis of the assay (for example, K_1 and S). Although several methods can be suggested for using the preceding analysis to extract information related to the response (DeLisi, 1975c), their application generally involves tedious experimentation. They are, in addition, relatively restricted compared to the inhibition techniques to be developed in the next chapter. I will therefore defer discussion of response parameter estimation and consider some general characteristics of the equations and their relation to experiment.

The possible growth patterns of plaques are clearly going to depend upon the range of values accessible to the control parameters. First consider k_2'. It will reflect epitope density and antibody class. In Chapter 2, a value of 2.1×10^5 sec^{-1} was obtained for IgM based on a guess of 100 DNP groups per phage. This density implies an average distance of about 50 Å between epitopes which is to be compared to a mean combining site separation of 87 Å for an antibody with a fully flexible hinge. If these estimates are reasonably accurate, it would appear that the phage can be considered densely haptenated. For an RBC with a diameter of about 7×10^{-4} cm, the same mean epitope spacing would require about 8×10^6 groups, about an order of magnitude larger than the estimated number of Forssman epitopes (Humphrey, 1967) and probably close to an experimental upper limit.

As epitope density decreases, k_2' will of course also decrease, initially as a linear function of E. However, as E continues to decrease, the behavior of k_2' will become considerably more complicated as the mean epitope spacing becomes larger than the distance between combining sites and diffusion becomes a requirement for intramolecular reaction.

Unfortunately, because detailed knowledge of the hinge is lacking and the necessary quantitative information on membrane characteristics is not available, a theoretical analysis of the behavior of k_2' is not possible. The problem is best studied experimentally and the results will most likely depend on antibody class as well as epitope density.

The other parameters which are important determinants of plaque growth are k_1, ρ_0 and K_1. For an antibody combining site interacting with free ligand, k_1 is typically about 10^7 $(M\text{-sec})^{-1}$. However, when the ligand is bound to a surface, binding constants several orders of magnitude smaller have been observed (Hughes-Jones et al., 1963; Jerne et al., 1974).

It is possible that in these latter experiments, k_1 is not being measured. As the analysis in Chapter 2 indicates, the observed association rate for two-step reactions is generally a complicated function of the rate constants for all the elementary steps. Only when the system is properly prepared will association be controlled by k_1. For the Hornick-Karush experiments, k_1 apparently was measured and its value turned out to be about 10^7 $(M\text{-sec})^{-1}$, essentially the same as the forward rate constant for free ligand. This suggests that there is little inherent difference in k_1 for binding aggregated (i.e., bound to a sur-face) rather than free ligand. I will therefore take 10^7 $(M\text{-sec})^{-1} = 1.6 \times 10^{-14}$ (molecules/sec-cm^3) as a typical forward rate constant.

The rate of bond formation of course depends on both the forward rate constant and the concentration of reacting units. Concentrations of about 4×10^8 RBC/cm^3 are typical of many plaque experiments and E may range from $10^3 - 10^6$. Thus ρ_0 ranges from $(4 \times 10^{11} - 4 \times 10^{14})$ epitopes/cm^3, and $k_1 E$ from $(1.6 \times 10^{-11} - 1.6 \times 10^{-8})$ sec^{-1}.

One effect of parameter variation can be seen in equation 5.23 according to which D_0 is reduced by a factor $\dfrac{1}{1+2K_1\rho_0}$. This reflects the hindrance of antibody motion brought about by the presence of specifically reactive RBC binding sites. In the absence of intra-molecular reaction, free antibody diffusion is interrupted by occasional, momentary encoun-ter with epitope. The magnitude of the effect is determined by the magnitude of $2K_1\rho_0$. If binding is tight, hindrance may be severe; whereas if it is weak, diffusion is relatively unimpeded. Thus, for the upper limit on ρ_0, unhindered diffusion requires $K_1 < 7.5 \times 10^5$ M^{-1} whereas for the upper limit, $K_1 < 7.5 \times 10^8$ M^{-1} is sufficient. Evidently it is not difficult to achieve conditions under which hindrance plays a relatively minor role. In such circumstances the interpretation of plaque size simplifies considerably since the only unkown parameter in equation 5.24 is $(K_1 S_0)$ and therefore the plaque size distribution will reflect the $(K_1 S_0)$ distribution.

5.6.2   Plaque Growth Characteristics.   Comparison With Experiment

In two dimensions, the local equilibrium limit leads to a very simple relationship between plaque area and time. Since the left side of equation 5.30 is constant, the argument of the exponential integral must also be constant. Consequently, if local equilibrium conditions are met, equation 5.30 predicts that in a two-dimensional system (Cunningham 1965), the plaque radius grows as the square root of time. Equivalently, if the plaque radius does not grow as the square root of time, local equilibrium conditions do not prevail. The converse of course need not be true; i.e., square root of time growth does not necessarily imply an equilibrium-controlled reaction. In experiments with IgM, Ingraham and Bussard (1964) observed a linear relationship between plaque area and time, so local equilibirum may have prevailed for their conditions. This possibility will be discussed Chapter 7 along with other pertinent data. For now, however, it is sufficient to indicate the relationship.

For a three-dimensional system equation 5.24 predicts that a change in either the total epitope concentration (ρ_0) or the number of epitopes per RBC (E) will effect plaque growth. With regard to the former, keeping E fixed and increasing the RBC concentration will decrease D and hence reduce r_p. This is what is seen experimentally as the data of Jerne et al. (1974) indicate (Figure 5.1). It should also be noticed, however, that the number of plaques is not reduced substantially. That is because the first plaques to disappear are the smallest and, according to Figure 5.1, these are the least abundant.

The relatively low number of small plaques can be understood in terms of the th According to equation 5.24, the largest plaques are associated with neither the lowest nor the highest affinity antibodies, but with some intermediate value. The reason is that K_1 enters the expression for r_p in two opposing ways. As K_1 increases, the chance of binding and, hence, of RBC lysis increases and this effect tends to increase r_p. However, because high affinity antibodies bind more tightly, they do not diffuse as far. This is reflected in a reduced diffusion coefficient and hence a smaller value of the co-error function

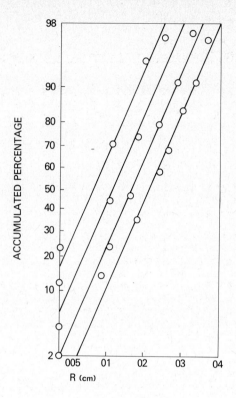

Figure 5.1. The accumulated percentage of plaques, on a probit scale, having radius R at four different RBC concentrations. The line furthest right corresponds to the lowest concentration.

This means that the smallest plaques are characterized by the lowest and highest affinity antibodies and, therefore, fall at the extreme low amplitude end of the affinity distribution. Consequently, as the RBC concentration is increased, the smallest and least abundant plaques disappear first so that over some range in concentration there is not a substantial reduction in their number.

The above remarks apply to changes in the total epitope concentration, ρ_0. It is also possible to vary E, keeping ρ_0 constant. This can be done, for example, by increasing the number of epitopes per RBC while simultaneously reducing the number of RBCs used in the assay. Alternatively, E can be varied with no compensation in the RBC concentration thus also affecting ρ_0.

Pasanen and Mäkelä (1969) have performed experiments conforming to the latter conditions, as a function of time after immunization. Early in the immune response whe

antibodies are of relatively low affinity (Chapter 3), large numbers of IgG plaques were

obtained with densely haptenated RBCs, but none appeared on lawns prepared with cells

low epitope density. The difference may be interpreted as a switch from conditions und

which equation 5.21 is applicable to those under which equation 5.24 applies. In the fo

case there is intramolecular bonding (high E) so even though K_1 is small, the extra bor

may be responsible for a relatively large effective equilibrium constant, and hence a su

stantial concentration of bound antibody (Chapter 2). In the latter case only univalent

attachment is possible (low epitope density) so it is not surprising to find no plaques fo

a low affinity population.

For antibodies produced late in the response (relatively high affinity), plaques

were produced even at the lowest epitope density. However, with increasing E, the nu

of observed plaques at first increased, but then began to decrease. There is unfortuna

no unique explanation for this observation without knowledge of the experimental value

E (which were not determined) and the behavior of k_2'. The general pattern, however,

clearly within the range of possibilities predicted by the theory. For example, one pos

bility is that at the lowest values of E, equation 5.24 holds and a relatively small numbe

plaques is seen. As E increases and intramolecular bonding starts to become important

more antibodies are bound and consequently the observed number of plaques increases

But as E continues to increase, L becomes very small. This means that many antibodie

are absorbed by only the few closest RBCs producing insufficient lysis to see a plaque.

Thus there will be preferential inhibition of plaques produced against low secretion rat

and/or high affinity antibodies. This general behavioral pattern has also been observe

by Inman et al. (1973).

6

PLAQUE INHIBITION: GROWTH KINETICS IN THE PRESENCE OF COMPETITIVE INTERACTIONS

.1   Introduction

A quantitative understanding of the dependence of antibody affinity and secretion
ate distributions on immune parameters may provide considerable insight into the dynamics
f the response. However, it is precisely heterogeneity in the parameters and its diverse
rigins which complicate the analysis of experiments. The problem is illustrated by the re-
ults of the last three chapters. When antiserum is used as a source of antibodies, the po-
ential information is relatively coarse (see Chapter 3) and the methods for obtaining it still
eed to be developed rigorously. When lymphocytes are the source, the complications of
eterogeneity vanish, but characterizing the evolution and dynamics of the response distri-
utions then requires studying the growth patterns of large numbers of plaques individually,
n enormously burdensome task. These difficulties can be circumvented by the plaque
hibition technique.

Inhibition experiments are generally used to analyze responses to hapten carrier
njugates and therefore the RBCs in the assay have the immunizing hapten covalently cou-
led to their membranes. However, free ligand (sometimes referred to as inhibitor) is also
resent in the mixture and since it is chemically identical to the epitope against which the
ntibodies are directed, it competes for combining sites. The concentration of free antibody
tes available for binding RBCs is thus reduced, leading to a reduction in the concentration
RBC-bound antibodies, and consequently in lysis, plaque size and plaque number.

The effectiveness of the competition depends on the ratio of inhibitor to epitope con-
ntration. Thus if a series of petri dishes is prepared with a different inhibitor concentra-
on in each, the dish with the highest concentration will show the smallest number of plaques
greatest extent inhibition), and the number will increase as concentration decreases. A
ot of the number of plaques as a function of inhibitor concentration is referred to as an
hibition curve; and a plot of the slope of this curve against inhibitor concentration is a
fferential plot. It is generally believed that the form of a differential plot reflects the af-
nity distribution, and many biological conclusions have been based upon this association.

In order to understand how a differential plot might reflect the affinity distribu
it is useful to think of the antibody population as a superposition of discrete non-overlap
subpopulations, each characterized by a particular affinity. Then the usual argument is t
at sufficiently low inhibitor concentrations, only the highest affinity antibodies will be blo
and the corresponding plaques inhibited. In a plate with a somewhat higher concentrat
the highest affinity subpopulation will still be blocked, but now there is sufficient inhib
so that the next lowest subpopulation will also be bound appreciably. Thus a greater n
ber of plaques will be inhibited, and so on with increasingly high concentrations until
plaques have disappeared. To put this in a slightly more formal way: If H is the inhib
concentration at which the extent of inhibition is P $(0 \le P \le 1)$, then a variation, ΔH, le
to a variation in the extent of inhibition, ΔP, and the magnitude of ΔP should be propor
to the number of antibodies in the additional subpopulation that is inhibited. Consequel
it has been proposed that: (a) a plot of the slope of the inhibition curve against inhibit
concentration will reflect the antibody affinity distribution (Wu and Cinader, 1971; Davi
and Paul, 1972; Clafin, Merchant and Inman, 1973) and (b) a measure of the relative af
nities of two antibody populations is provided by comparing the amount of inhibitor need
to achieve 50% (say) inhibition (Andersson, 1970).

This reasoning involves some implicit and non-transparent assumptions about
physical chemical conditions of the experiment. My purpose is to develop these explic
within the framework of a mathematical model of inhibition, and to define the condition
which must be met in order to obtain affinity or secretion rate information. In Chapter
the theory will be applied to the analysis of the appropriate experimental data, and the b
logical implications of the results developed.

6.2   IgG Plaque Inhibition

6.2.1   A Generalized Diffusion-Reaction Model

As antibody diffuses from its source, inhibitor molecules may bind one or both
its sites. When this happens, the antibody continues to diffuse, though with a somewh
different diffusion coefficient. If the inhibitor is a small hapten such as DNP, the chan
may be negligible; if it is large, such as $DNP_{E'}$ - BSA (bovine serum albumin with E' I
groups per molecule), the change may be substantial. More importantly, however, th

binding reaction partitions the antibodies into three distinct physical chemical species.

Those with both sites blocked cannot bind RBCs. Those with one blocked may bind only

univalently, and hence have very rapid dissociation rates, while those with both free can

bind bivalently and essentially irreversibly. I will denote the concentrations of antibodies

with inhibitor binding a single site by C_3 or C_4, according to whether the other site has

bound to RBC or is free. In the latter case the antibody is diffusing whereas in the former

it is not. In addition, C_5 and C_6 will denote the concentrations of diffusing antibodies with

both sites blocked; the former denoting complexes with two inhibitor molecules and the

latter with one (Figure 6.1).

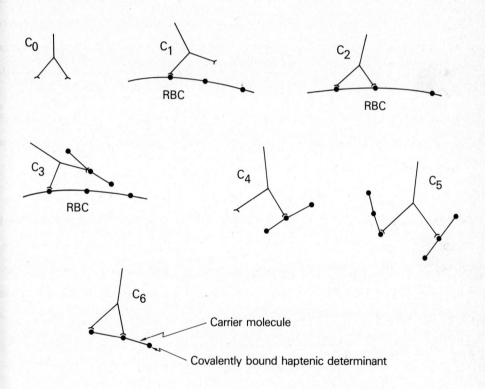

Figure 6.1. The C_i are the concentrations of antibody in the states indicated.

With these definitions, the equations of the model can be written in the following

form

$$\frac{\partial \vec{C}}{\partial t} = \vec{D} \cdot (\nabla^2 \vec{C}) + \underset{\sim}{M} \vec{C}$$

$$6.1$$

where

$$\vec{C} = (C_0, C_1, \ldots C_6)^\tau$$

τ denoting the transpose,

$$\vec{D} = (D_0, 0, 0, 0, D'_0\ D''_0\ D'_0)$$

and (Figure 6.2)

$$
M = \begin{pmatrix}
-2k_1(H+\rho_0) & k_{-1} & 0 & 0 & k_{-1} & 0 & 0 \\
2k_1\rho_0 & -(k_{-1}+k_1H+k'_2) & 2k_{-2} & k_{-1} & 0 & 0 & 0 \\
0 & k'_2 & -2k_{-2} & 0 & 0 & 0 & 0 \\
0 & k_1H & 0 & -2k_{-1} & k_1\rho_0 & 0 & 0 \\
2k_1H & 0 & 0 & k_{-1} & -(k_{-1}+k_1H+k_1\rho_0+k''_2) & 2k_{-1} & 2k_{-2} \\
0 & 0 & 0 & 0 & k_1H & -2k_{-1} & 0 \\
0 & 0 & 0 & 0 & k''_2 & 0 & -2k_{-2}
\end{pmatrix}
$$

D'_0 and D''_0 denote the diffusion coefficient of antibodies complexed to one and two inhibitor molecules, respectively, k''_2 is the rate constant for establishing an intramolecula[r] bond between antibody and multifunctional inhibitor, and H is the free inhibitor concen[tra]tion. Since H is always very large compared to the antibody concentration, it is essent[ially] constant and equal to the total inhibitor concentration. These equations follow at once [from] the chemical reaction scheme in Figure 6.2 and the definitions in Figure 6.1, keeping i[n] mind that only species C_0, C_4, C_5, and C_6 diffuse.

It will be convenient to keep track of the total concentration of diffusing antibo[dy] C, which is defined as (Figure 6.1)

$$C \equiv C_0 + C_4 + C_5 + C_6$$

6.[?]

From equations 6.1-6.5 it follows that

$$\frac{\partial C}{\partial t} = D_0 \nabla^2 C_0 + D_0' \nabla^2 C_4 + D_0'' \nabla^2 C_5 + D_0' \nabla^2 C_6 - \frac{\partial}{\partial t}(C_1 + C_2 + C_3)$$ 6.6

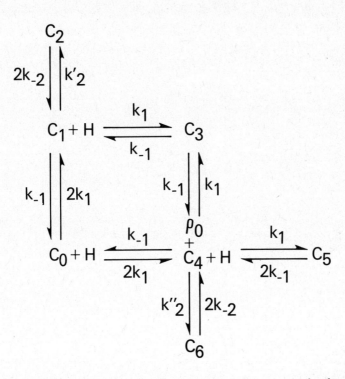

Figure 6.2. The chemical reaction scheme. k_2' is the rate constant for forming an intra-molecular bond with RBC; k_2'' the rate constant for forming an intramolecular bond with carrier.

Although the general model is relatively complicated, the physical considerations introduced earlier based upon the time scales of the processes lead to considerable simplification. In particular, I assumed that: (a) bivalent antibody binding is irreversible and (b) the time rate of change of singly bound antibodies (i.e., those which have not undergone intramo-lecular reaction) due to chemical reaction is close to zero, either for equilibrium reasons (sparse epitope density) or steady state reasons (very high epitope density). With these simplifications one finds that

$$C_1 \simeq \frac{2K_1 P_0 F_1 C_0}{F_1 + F_2}$$ 6.7

$$C_3 \stackrel{\sim}{=} 2K_1 H [J (1 + K_1 \rho_0 + K_2''/k_{-1}) - 1] C_0 \qquad 6.8$$

$$C_4 \stackrel{\sim}{=} 2K_1 H J C_0 \qquad 6.9$$

$$C_5 \stackrel{\sim}{=} (K_1 H)^2 J C_0 \qquad 6.10$$

where

$$J \equiv \left\{ 1 + \frac{K_1 \rho_0 [F_1 + (k_2' - k_2'')/k_{-1}]}{F_1 + F_2} \right\} \bigg/ (1 + K_1 \rho_0 + k_2''/k_{-1}) \qquad 6.11$$

$$F_1 \equiv 2 + K_1 (H + \rho_0) + 2k_2''/k_{-1}$$

and

$$F_2 \equiv 2k_2'/k_{-1} + K_1 Hk_2''/k_{-1} + K_1 \rho_0 k_2'/k_{-1} + 2k_2' k_2'' /k_{-1}^2$$

Some specific examples will now be developed.

6.2.2   Inhibition by Univalent Molecules

When the inhibiting ligand is a small molecule with only one combining site, C_6
Therefore defining Z as the sum of the weights of all concentrations relative to that of C
which is given weight 1, one has that

$$C = C_0 + C_4 + C_5 = C_0 Z \qquad 6.12$$

where from equations 6.9 and 6.10

$$Z = 1 + 2K_1 H J + (K_1 H)^2 J \qquad 6.13$$

Consequently

$$C_0 = C/Z \qquad 6.14$$

$$C_4 = 2K_1 H J C/Z \qquad 6.15$$

and

$$C_5 = (K_1 H)^2 J C/Z \qquad 6.16$$

Therefore equation 6.6 becomes

$$\frac{\partial C}{\partial t} = D^* \nabla^2 C - \frac{\partial}{\partial t}(C_1 + C_2 + C_3) \qquad 6.17$$

where

$$D^* \equiv (D_0 + 2K_1 H J D_0' + (K_1 H)^2 J D_0'')/Z \qquad 6.18$$

According to the assumptions made in deriving equations 6.7 and 6.8, the derivatives of C_1 and C_3 in equation 6.17 are non-zero only because C is changing by diffusion. Thus from equations 6.7, 6.8 and 6.14

$$\frac{\partial}{\partial t}(C_1 + C_3) = \frac{2K_1 P_0}{Z}\left[\frac{F_1(1 + K_1 H) + K_1 H k_2''/k_{-1}}{F_1 + F_2}\right]\frac{\partial C}{\partial t} \qquad 6.19$$

In addition

$$\frac{\partial C_2}{\partial t} = k_2' C_1 \qquad 6.20$$

which, according to equations 6.4 and 6.7, becomes

$$\frac{\partial C_2}{\partial t} = \frac{2K_1 P_0 F_1 k_2' C}{Z(F_1 + F_2)} \qquad 6.21$$

Substituting equations 6.19 and 6.21 into equation 6.17, one has that

$$\frac{1}{D}\frac{\partial C}{\partial t} = \nabla^2 C - \frac{C}{L^2} \qquad 6.22$$

where

$$D \equiv \frac{D^*}{1 + \frac{2K_1 P_0}{Z}\left[\frac{F_1(1 + K_1 H) + K_1' H k_2'/k_{-1}}{F_1 + F_2}\right]} \qquad 6.23$$

and

$$L^2 \equiv \frac{D^* Z (F_1 + F_2)}{2K_1 P_0 F_1 k_2'} \qquad 6.24$$

Thus the equation describing total antibody diffusion has precisely the same form as equation 6.10, only with the effective diffusion coefficient and diffusion length altered. The formal similarity is not surprising since the physical assumptions used in deriving them are the same. The effect of inhibitor is only to change the time and distance scales of the proce

Hence equation 5.16 is the solution to equation 6.22 with a continuous time invariant so

and with D and L given by equations 6.23 and 6.24, respectively. The quantity of expe

mental interest, the plaque radius, is given by the solution to equation 6.25.

$$N^* = \frac{2K_1 E}{Z(F_1 + F_2)} \left\{ F_1 C + K_1 H (F_1 + k_2'/k_{-1})C + k_2'F_1 \int_0^t C(r_p,t') \, dt' \right\} \qquad 6.2$$

where $C(r,t)$ is the solution to equation 6.22. As intramolecular reaction becomes eithe

instantaneous or infinitesimally slow, the following limiting solutions hold.

Steady state: $\dfrac{2k_{-1} + k_1 H}{k_2'} \to 0$

$$N^* = \frac{EC}{2H} + \frac{k_1 E}{2K_1 H} \int_0^t C(r_p,t') \, dt' \qquad 6.2$$

where $C(r,t)$ is still the solution to equation 6.22 but with

$$D^*/(1 + \rho_0/H) \overset{\sim}{=} D^* \qquad D^* = D_0 + \frac{4K_1 H D_0'}{2 + K_1\rho_0} + \frac{2(K_1 H)^2 D_0''}{2 + K_1 \rho_0} \qquad 6.2$$

$$L^2 = \frac{D^* K_1 H}{k_1 \rho_0} \qquad 6.2$$

and use has been made of the fact that within the inhibition zone $H > \rho_0$ and $K_1 H > 1$ (L

and Goldstein, 1975).

Equilibrium: $\dfrac{2k_2'}{2k_{-1} + k_1 H} \to 0$

The plaque radius is given by

$$r_p = \frac{SK_1 E}{2\pi(1 + K_1 H)D_0 N^*} \text{ erfc} \left(\frac{r_p}{2\sqrt{Dt}} \right) \qquad 6.29$$

with

$$D = D^*/(1 + 2\rho_0/H) \qquad\qquad 6.30$$

and L as infinite.

6.2.3   Inhibition by Multivalent Molecules

In this case inhibition is achieved by using a large carrier molecule such as a protein with E' haptens covalently bound to its surface. Thus the total inhibitor concentration, H, is E' times the carrier concentration. Typically, the number of haptens per carrier corresponds to relatively high densities. For example, since the radius of BSA is about 10^{-3} that of the RBC, ten DNP groups on the former would correspond to about ten million on the latter. It is therefore reasonable (or at least consistent) to require irreversible binding of antibody to inhibitors when epitope densities are in this range. However, when there are very few epitopes per carrier, an intramolecular reaction may require considerable strain in the hinge as the result of a high degree of inhibitor curvature (e.g., the BSA surface cannot be considered flat on the scale of antibody dimensions). Under such circumstances K_2 may be small due to an inordinately high reverse rate constant. I will therefore develop two limiting cases: one in which epitope densities may be considered high on both RBC and carrier, and another in which they are low on both.

High RBC Epitope Density $\left(\dfrac{k_2'}{k_{-1}} \to \infty ; \dfrac{k_2''}{k_{-1}} \to \infty \right)$

With intramolecular bond formation very rapid compared to dissociation, C_4 and C_1 are assumed to be in local steady states and at concentrations which are entirely negligible compared to those of competing species. For the same reason, the concentrations, C_5 and C_3, must also be negligible. Thus, with the bimolecular step rate limiting the formation of bivalently bound complexes, and with the second bond established instantaneously,

$$C = C_0 + C_6 \qquad\qquad 6.31$$

and equation 6.6 is approximated by

$$\frac{\partial C}{\partial t} = D_0 \, \nabla^2 \, C_0 + D_0' \, \nabla^2 \, C_6 - \frac{\partial C_2}{\partial t} \qquad\qquad 6.32$$

In addition

$$\frac{\partial C_6}{\partial t} = D_0' \nabla^2 C_6 + k_2'' C_4 \qquad\qquad 6.33$$

C_6 is essentially inert, so there is no reason to keep explicit account of it. The diffusion equation of interest is the one governing C_0. Subtracting equation 6.33 from 6.32 and using 6.31

$$\frac{\partial C_0}{\partial t} = D_0 \nabla^2 C_0 - \frac{\partial C_2}{\partial t} - k_2'' C_4 \qquad\qquad 6.34$$

where (equation 6.1)

$$\frac{\partial C_2}{\partial t} = k_2' C_1 \qquad\qquad 6.35$$

Substituting equation 6.7 into equation 6.35 and letting k_2'/k_{-1} and k_2''/k_{-1} become infinite

$$\frac{\partial C_2}{\partial t} = 2k_1 \rho_0 C_0 \qquad\qquad 6.36$$

Using equations 6.9 and 6.11 to evaluate $k_2'' C_4$ in the same way, one finds that (DeLisi 1975a)

$$\frac{\partial C_0}{\partial t} = D_0 \nabla^2 C_0 - 2k_1 (H + \rho_0) C_0 \qquad\qquad 6.37$$

Equation 6.37 has a solution of the form of equation 5.13 with

$$D = D_0 \qquad\qquad 6.38$$

and

$$L^2 = \frac{D_0}{2k_1 (H + \rho_0)} \qquad\qquad 6.39$$

The number of antibodies per RBC at (r,t) is obtained by integrating equation 6.36 and dividing by the RBC concentration.

$$N(r,t) = 2k_1 E \int_0^t C_0 (r,t') dt' \qquad\qquad 6.40$$

Consequently the plaque radius is found by solving

$$N^* = 2k_1 E \int_0^t C_0 (r_p, t') dt' \qquad 6.41$$

Sparse Haptenation on Both RBC and Carrier. For reasons which will be discussed in Chapter 7, I will consider the limit in which attachment to RBC is univalent, and attachment to inhibitor bivalent but strained. Because of strain, the intramolecular reaction can no longer be considered irreversible and, in fact, k_{-2} may be so large that equilibrium is quickly established. Therefore the concentrations C_0, C_4, C_5 and C_6 will be assumed to change by diffusion alone, i.e., their first time-derivatives due to chemical reaction vanish With $k_2'/k_{-1} \rightarrow 0$ (sparse RBC epitope density), one has that

$$C_1 = 2K_1 \rho_0 C_0 \qquad 6.41a$$

$$C_3 = 2K_1 \rho_0 K_1 H C_0 \qquad 6.41b$$

$$C_4 = 2K_1 H C_0 \qquad 6.41c$$

$$C_5 = (K_1 H)^2 C_0 \qquad 6.41d$$

$$C_6 = K_2'' K_1 H C_0 \qquad 6.41e$$

and equation 6.5 for the concentration of diffusing antibody molecules becomes

$$C = C_0 (2K_1 \rho_0 + 2K_1 H + (K_1 H)^2 + K_2'' K_1 H) \equiv C_0 Z \qquad 6.42$$

Substituting 6.41 and 6.42 into 6.6

$$\frac{\partial C}{\partial t} = D \nabla^2 C \qquad 6.43$$

where

$$D = \frac{D_0 + D_0' (2K_1 H + K_2' K_1 H) + D_0'' (K_1 H)^2}{Z + 2K_1 \rho_0 (1 + K_1 H)} \qquad 6.44$$

From 6.41a, 6.41b and 6.42 the number of bound antibodies per RBC is

$$N = \frac{2K_1 E (1 + K_1 H) C}{Z} \qquad 6.45$$

where C is the solution to equation 6.43. More specifically, at the plaque boundary

$$N^* = \frac{2K_1 E \ (1 + K_1 H) \ S_0}{4\pi \ D_0 \ Z \ r_p} \ \text{erfc}\left(\frac{r_p}{2\sqrt{Dt}}\right) \qquad 6.46$$

where D is given by equation 6.44, and Z by 6.42.

6.3 Predictions

The question of interest is: if the experimental system were to meet the conditi of the derivation, which parameter(s) would a differential plot reflect? In order to obta an answer, inhibition must be defined more precisely. As I will show shortly, an increas in H leads to a reduction in plaque radius. Eventually the plaque becomes so small that i cannot be reliably distinguished from fluctuations in the graininess of the RBC lawn, an it is in essence inhibited. This happens at a plaque radius of about 15 or 20 μ. To be s ific, I will assume 20 μ (DeLisi and Goldstein, 1975), but the precise value will not affe the qualitative conclusions.

Univalent Inhibitor. As an example of what is to be expected, suppose $n(K_1)$, fraction of cells secreting antibodies of affinity K_1, is known (Table I). Then the theor

TABLE I

HYPOTHETICAL AFFINITY DISTRIBUTION

K_1 (cm^3/molecule)	$n(K_1)$
1.7×10^{-11}	0.0044
8.45×10^{-12}	0.0700
4.20×10^{-12}	0.121
2.1×10^{-12}	0.1590
1.05×10^{-12}	0.1400
5.25×10^{-13}	0.1100
2.60×10^{-13}	0.0700
1.30×10^{-13}	0.0150
6.50×10^{-14}	0.0460
3.25×10^{-14}	0.0840
1.62×10^{-14}	0.1100
8.10×10^{-15}	0.0540
4.10×10^{-15}	0.0166

can be used to simulate an experiment and predict a differential inhibition plot which can be compared to $n(K_1)$. This is done by using either equation 6.26 or equation 6.29 to determine, for each subpopulation, the inhibitor concentration at which r_p is reduced to 20 µ. At each value of H, one therefore knows which subpopulations are inhibited and hence the fraction of plaques remaining.

The inhibition curve predicted under the assumption that antibody binds irreversibly to the RBC (equation 6.26) is shown in Figure 6.3. In Figure 6.4 its derivative is compared with the affinity distribution. The striking similarity between $n(K_1)$ and the differential plot supports intuitive expectations. However, it must be emphasized that rather stringent conditions were imposed on the derivation, which may be difficult to meet experimentally. Before discussing these, it is worth considering the reason for the result.

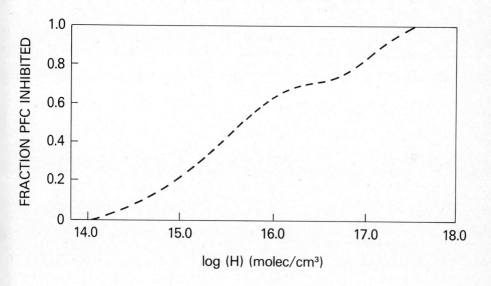

Figure 6.3

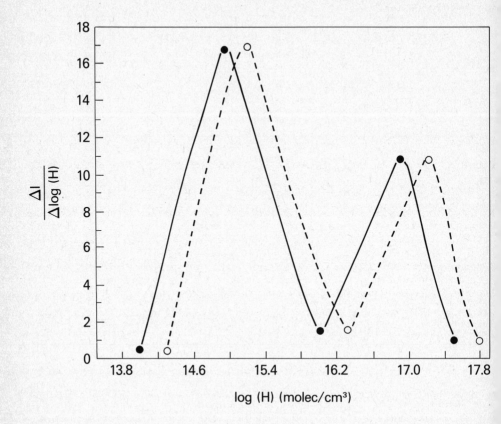

Figure 6.4. ΔI is the fraction of PFCs inhibited in the concentration interval $\Delta \log (H)$. -●- is the differential plot; --O-- the form of the affinity distribution given in Table I.

The similarity between the differential plot and the affinity distribution can be understood by first removing from equation 6.26 terms which are not important under conditions prevailing in plaque experiments. Specifically, there are two simplifications. In order to inhibit all but the very smallest plaques, $K_1 H \gg 1$. In addition, since k_{-1} sec^{-1} and $C(20\,\mu, t)$ is considerably less than its integral, the first term on the right of tion 6.26 can be neglected. Thus the equation describing the boundary of the smallest servable plaque $(r_p = r_{min} \cong 20\,\mu)$ is

$$N^* = \frac{k_1\, ES}{K_1 H} \int_0^t C'\, (r_{min},\, t')\, dt' \qquad 6.47$$

where I have factored an S out of $C(r,t)$ given by equation 5.16 to obtain $C'(r,t)$. Under the above conditions

$$D \overset{\sim}{=} D_0'' \qquad\qquad 6.48$$

and

$$L^2 \overset{\sim}{=} \frac{D_0'' K_1 H}{k_1 \rho_0} \qquad\qquad 6.49$$

Evidently the affinity and inhibitor concentration enter equation 6.47 only as a simple product $K_1 H$. Since N^* is constant, changes in H must be compensated by changes in K_1. Thus an incremental increase in H leads to inhibition (i.e., a reduction of r_p to 20 μ) of the next lowest affinity subpopulation and a differential plot will therefore reflect the distribution.

The most important omission from this analysis is the secretion rate distribution. A more complete simulation of the experiment would involve generating inhibition curves for different secretion rates, weighting each by the amplitude of the distribution, and summing. The extent to which the resulting differential plot reflects $n(K_1)$ will depend on the form of the secretion rate distribution.

An example of what might be expected is shown in Figure 6.5. The curve was constructed by superimposing three differential plots with S = (100, 500 and 1,000) antibodies/ sec, after assigning weights of 0.2, 0.6 and 0.2, respectively. Although there is some distortion, the result is still a reasonable reflection of $n(K_1)$. However, more complicated secretion rate distributions could easily result in more complicated differential plots. Moreover, one cannot rule out, a priori, a correlation between affinity and secretion rate. Under these more general conditions, the relation between $n(K_1)$ and the differential plot will be more complicated, even for the limiting conditions leading to equation 6.26.

It should be noticed that the relationship shown in Figure 6.4 is obtained under initial conditions (i.e., H = 0) for which neither affinity nor secretion rate variation is possible. Since these are the two quantities which contribute to variation in plaque radius, that means all plaques must be the same size when inhibitor is absent. This result should not be entirely unexpected since, if there were an initial plaque size distribution, the smallest plaques would tend to be inhibited first. Under such circumstances the smallest plaques would have to be those of both the highest affinity and lowest secretion rate in order to find

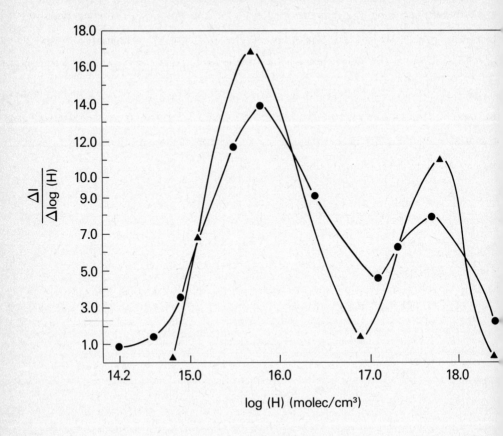

Figure 6.5. Comparison of the results obtained using a single secretion rate (-▲-) with those obtained using a secretion rate distribution (-●-).

a good relationship between $n(K_1)$ and the differential plot. However, it is evident from the results of the last chapter (see also DeLisi, 1976) that, in general, there is no simple relationship between plaque size and antibody affinity (when H = 0). For very low affinity populations, the largest plaques will tend to be those associated with the high affinity component of the population. The reason is that at large distances the antibody concentration is relatively low so that only those of relatively high affinity will bind sufficiently to effect lysis. On the other hand, for high affinity populations, hindered diffusion becomes an important effect. Most high affinity antibodies are held in close to the lymphocyte and the smallest plaques do tend to center on lymphocytes secreting the highest affinity anti-bodies. This argument suggests that deviation from the irreversible binding condition will be more important for low rather than high affinity populations. Mathematical investigation of the sensitivity of the results to variations from irreversible binding would be worthwhile.

The above analysis applies to an irreversible binding process rate limited by the bimolecular step. The other limit of interest is the equilibrium condition. In that case, equation 6.46 holds. From what was just said, it is evident that during an inhibition experiment, this equation can be adequately expressed by

$$r_{min} = \frac{2ES_0}{4\pi D_0 N^*H} \; erfc \left(\frac{r_{min}}{2\sqrt{D_0'' t}} \right) \qquad\qquad 6.50$$

This expression is independent of K_1, and for fixed S_0 there will be just one value of H at which r_{min} is reached. Consequently, inhibition is predicted to occur abruptly, as shown in Figure 6.6.

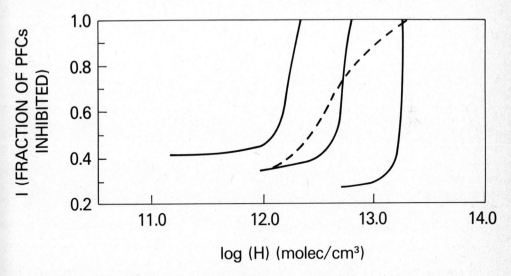

Figure 6.6. The solid lines are results obtained for S = 100, 500 and 1,000, reading from left to right. The dashed line is a superposition of the three single secretion rate results.

For the more realistic case in which a distribution of secretion rates is allowed, the differential plot is broad. Rather than reflecting affinity, it now reflects secretion r_ Thus, depending upon RBC hapten density, a differential plot may reflect either the se- cretion rate or the affinity distribution.

It is easy to express these numerical results in simple algebraic form. Let N_p de note the number of plaques remaining at inhibitor concentration H, and let $n(K_1, S)\, dK_1$ be the fraction of the total population of ASCs whose secretion rate is between S and S+dS and whose antibodies have affinity for hapten between K_1 and K_1+dK_1. According to equ tion 6.50, all plaques with $S_0 \leq S_0^*$ will be inhibited, where

$$S^* = \frac{H}{\theta^*} \qquad\qquad 6.51$$

where

$$\theta^* \equiv \frac{E\ \mathrm{erfc}\ (r_{min}/2\sqrt{D_0''\,t})}{2\pi D_0 N^* r_{min}} \qquad\qquad 6.52$$

Therefore

$$N_p = \int_{K_{min}}^{K_{max}} \int_{S^*}^{S_{max}} n(K,S)\, dK\, dS \qquad\qquad 6.53$$

Differentiating equation 6.53,

$$\frac{dN_p}{dH} = - \int_{K_{min}}^{K_{max}} n(K,S^*)\, \frac{dS^*}{dH}\, dK \qquad\qquad 6.54$$

But from equation 6.51,

$$\frac{dS^*}{dH} = \frac{1}{\theta^*} \qquad\qquad 6.55$$

Therefore

$$\frac{dN_p}{dH} = -\frac{1}{\theta^*} \int_{K_{min}}^{K_{max}} n(K,S^*) \, dK \equiv -\frac{1}{\theta^*} g(S^*) \qquad\qquad 6.56$$

Equation 6.56 says that the slope of the inhibition curve is proportional to the amplitude of the secretion rate distribution at S^*, precisely what is illustrated in Figure 6.6.

Multivalent Inhibitor. In this case, interaction between antibody and inhibitor is assumed bivalent, irreversible and rate limited by the bimolecular step. The equation of interest is 6.41.

The predicted results when binding to both RBC and inhibitor is irreversible are shown in Figure 6.7. The curves are sensitive to neither affinity nor secretion rate. They drop precipitously, and the total inhibition is achieved over a very restricted concentration range. This is to be compared to the broad inhibition curves obtained when small hapten is used.

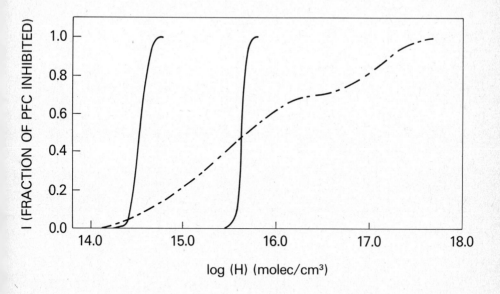

Figure 6.7. The solid lines are predicted inhibition curves for multivalent inhibitor. The displacement reflects an order of magnitude difference in the forward rate constants. For comparison, the dashed line shows the predicted result for monovalent inhibitor.

The difference between the models which lead to these results resides in the natu of the interaction with inhibitor. The reaction between antibody and monofunctional mole cules is equilibrium-controlled and leads to inhibition which is mediated by the term $\frac{1}{K_1 H}$ prefacing the integral in equation 6.47. Physically, this is the factor by which the concer tration of free sites on diffusing antibodies is reduced by the presence of hapten. Evidentl this must lead to an inhibition curve as broad as the affinity distribution since K_1 and H are reciprocally related on the plaque boundary.

By contrast, when the inhibitor is multifunctional, the reaction is affinity-indepen since the antibody cannot dissociate. This, however, is not the reason for abrupt inhibitio since it will be recalled that affinity-independent inhibition by monofunctional molecules leads to a relatively broad curve reflecting the secretion rate distribution. The reason th does not happen in the present case is that the concentration of free antibody sites depend exponentially (rather than linearly) on H. The physical origin of this is, in essence, that the fractional reduction in the antibody flux upon traversing a shell of perfect absorbers is proportional to the number of absorbers in (and hence the thickness of) the shell. In this respect the process is similar to many other absorption phenomena.

It should be noticed that with a fixed ρ_0 the position of the curve varies approxi- mately linearly with k_1. Thus the method provides some information about forward rate constants, both within and between populations. For example, a very narrow inhibition curve for a particular experiment clearly implies a very narrow forward rate constant di tribution. If two populations meeting these conditions are compared, and the inhibition curves are shifted, the separation will be proportional to the difference in forward rate constants between populations.

Finally, a change in ρ_0 will shift the position of the curve. This is because inhi tion depends on $\rho_0 + H$. Consequently, increasing ρ_0 leads to a smaller diffusion length and a smaller value of H needed to bring about inhibition.

7

APPLICATIONS OF PLAQUE GROWTH THEORY AND
IMMUNOLOGICAL IMPLICATIONS

7.1   The IgM Response

The ideas developed in the two previous chapters will now be applied to the analysis of hemolytic plaque data and the biological implications of the results discussed. Since
many interesting experiments have been performed with IgM as well as IgG and little has
been said of the former, I will begin by describing some general characteristics of its behavior in plaque experiments.

An accurate description of the thermodynamic and kinetic properties of IgM is difficult for a number of reasons. Part of the problem is complexity, and part inadequate knowledge of structural detail. The presence of additional combining sites introduces many more
possible states for the diffusing antibody, and many more ways in which it may interact with
an RBC. In addition, morphological variations are possible; i.e., the spatial disposition of
the five subunits relative to one another may shift. Moreover, conditions which favor com-
petitive morphologies are not known. Therefore, rather than attempt a priori formulations
of different models (DeLisi, 1975b), I will consider the results of some plaque inhibition experiments in order to place constraints on the possible range of IgM characteristics.

Figure 7.1 (Clafin, Merchant, Inman, 1973) compares experimental results obtained
using inhibitors of different valence. Inhibition is abrupt when univalent hapten is used
and broad when it is aggregated on a carrier (in this case, BSA). This is in striking contrast to IgG results. The obvious question is what this behavior might mean in terms of the
nature of the interaction between IgM and RBC.

There are only a limited number of possible interactions, and the consequences of
most are not compatible with what is known. To begin with, since monovalent inhibitor [in
this case 6-(2,4,6-trinitrophenylamino) caproic acid (TNP-EACA)] is somewhat smaller
than an antibody combining site, it cannot be bound bivalently (i.e., two antibody sites
cannot bind a single hapten). In keeping with previous reasoning, the interaction with
inhibitor is therefore equilibrium-controlled. Given this, the question is whether anything
can be deduced about the nature of the interaction with RBC.

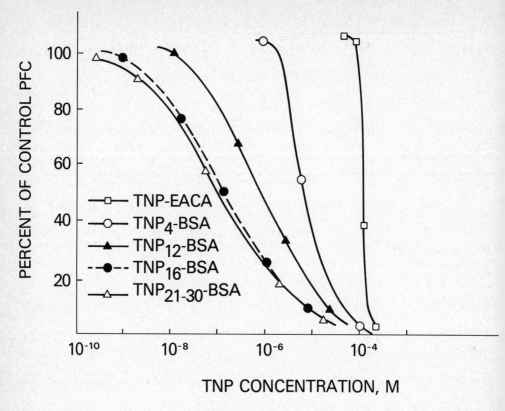

Figure 7.1

The results of Chapter 6 indicate that with the reaction between antibody and inh bitor equilibrium-controlled, irreversible binding of antibody to RBC will lead to an inhi- bition curve whose breadth (at fixed secretion rate) reflects the range in antibody affiniti Consequently, if the interaction with RBC were irreversible, the conclusion would be tha the IgM affinity and secretion rate distributions are very narrow. Although little is know about the latter distribution, the former spans several orders of magnitude. Thus irre- versible binding is unlikely.

The opposite assumption is that the reaction with RBC is equilibrium-controlled. In this case, with both the antibody-inhibitor and antibody-epitope interactions in local equilibrium, inhibition will not reflect the affinity distribution (Chapter 6). Thus a shar curve could occur under these conditions, provided the secretion rate distribution is na: row. This interpretation raises two obvious questions.

(1) Why should the interaction with RBC be equilibrium-controlled when, under identical conditions, the interaction between IgG and RBC is approximately irreversible?

(2) Why should the secretion rate distribution be narrow?

Some progress can be made toward answering the first question by considering the requirements for an equilibrium-controlled reaction between antibody and RBC. First, it seems likely that the rate for forming an intramolecular bond must be slow compared to dissociation of the monovalently bound antibody. Otherwise, multisite adherence would occur rapidly, and since such binding is irreversible within the time of the experiment, the equilibrium constant could not enter the description of the process. The structural implication is that the IgM hinge is somewhat less flexible than that of IgG.

A second structural implication is related to IgM morphology. Both staple and planar forms have been visualized (Feinstein et al., 1971). However, since the staple should lead relatively easily to multivalent binding, it would appear that this conformation is not favored. There is, in fact, some evidence which suggests that IgM free in solution has a planar form, and upon binding to a surface undergoes a conformational change to a staple-like morphology (Metzger, 1972).

Such behavior is consistent with the above reasoning, provided the time taken to change conformation and establish a second bond is long compared to the time required for monovalent dissociation. This would lead to a situation in which many associations and dissociations occur before the irreversible step. The effective forward rate constant for the process would therefore be proportional to K_1, the combining site-hapten equilibrium constant (see Chapter 2). It is also possible that complement binds IgM most efficiently when it is in the staple form (Metzger, 1974), a conjecture consistent with the above analysis.

The second major question related to the characteristics of IgM inhibition by monofunctional hapten involves the breadth of the secretion rate distribution. Since contact with antigen can lead to B cell differentiation, one would expect a response characterized by secretion rate heterogeneity. Consequently, secretion rate homogeneity suggests that up to the time of the experiment (4 days), little differentiation has occurred. B cell triggering may lead to an early proliferative phase, later followed by differentiation. Dutton (1975) has been led to a similar conclusion on the basis of other experimental evidence.

The sharp inhibition curve obtained with monofunctional hapten broadens as the inhibitor epitope density increases. Such broadening can be related to a gradual increase in the ease with which the antibody establishes multiple bonds. Provided intramolecular reaction is equilibrium-controlled, the breadth of the curve at high inhibitor epitope densities will reflect the breadth of the affinity distribution.

Although I will not develop the arguments in detail here, some insight into the conditions required for affinity-dependent inhibition is provided by equation 6.46. The result was obtained by assuming an equilibrium-controlled intramolecular reaction with inhibitor. It is evident that in this case, inhibition need not require large values of $K_1 H$. In particular, with $K_1 H < 1$ and $K_2'' K_1 H > 1$

$$N^* = \frac{E \, S_0 \, \mathrm{erfc}\left(r_p/2\sqrt{Dt}\right)}{2\pi \, D_0 \, r_p \, K_2'' \, H} = \frac{E \, S_0 \, \mathrm{erfc}\left(r_p/2\sqrt{Dt}\right)}{2\pi \, D_0 \, r_p \, A K_1 \, H} \qquad 7.1$$

where A is related to hinge flexibility and epitope density as discussed in Chapter 2 and is given by equation 6.44. The result predicts that $K_1 H$ is constant on the plaque boundary and hence changes in H are exactly compensated by changes in K_1. Since this is also true on the boundary of the minimum observable plaque, it follows (Chapter 6) that the breadth of the inhibition curve must reflect the breadth of the affinity distribution. Equation 7.1 also predicts that as epitope density increases (as reflected by an increase in A), H_{50} decreases, in agreement with observation (Figure 6.1).

If the above analysis is correct, information related to cellular selection mechanism may be obtained by studying inhibition as a function of time after immunization. The results of such experiments are shown in Figure 7.2 (Clafin et al., 1972). It is clear that as time increases, H_{50} decreases, implying a time-dependent affinity increase. The behavior appears to correlate reasonably well with changes in serum affinity, suggesting that the latter is at least in part attributable to pressure to select from the heterogeneous population of cells, those bearing high affinity receptors.

With increasing time after immunization, the slope of the inhibition curve sharpens and the midpoint shifts left, indication a narrowing affinity distribution and a preferential stimulation and dominance of clones secreting high affinity antibody. Moreover, the data indicate that the precursors of the dominant populations were present at day zero. This

important because it suggests that the full range of diversity is present prior to immunization. The alternative would be that some diversity is generated by mutation subsequent
to antigenic contact. Although this latter possibility appears biologically advantageous
since it would reduce the burden of having always to maintain the full range of specifici-
ties, the results tend to argue against it.

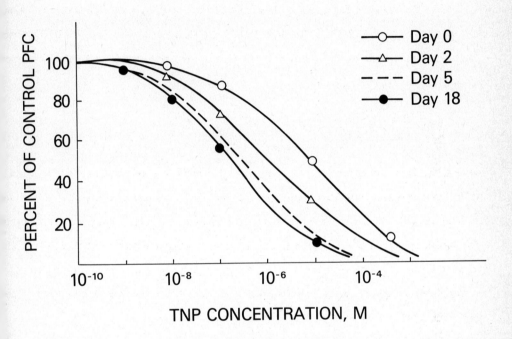

Figure 7.2

It is noteworthy that the major portion of the affinity change occurs within the first
five days after immunization, with the entire maturation process completed by day 9. In view
of what was said above, this means that most selection occurs during a proliferative phase
of the response. Thus the immune system appears to prefer preferential proliferation of cells
with high affinity receptors, followed by differentiation of the expanded clones into mature,
high secretion rate plasma cells.

Such a response also seems reasonable on teleologic grounds. For example, in an extreme situation with a single differentiation step leading to plasma cell formation, differ

entiation (rather than proliferation alone) of precursors would lead immediately to a popu

lation of cells which cannot divide and therefore could not mature. Consequently, allowin

only a single stage at which selection can occur abolishes the capacity for maturation, as

well as the capacity to mount a response with more than a minimum number of plaque-forr

ing cells. In a more general case, with several differentiation steps occurring before a

plasma cell is produced, there will be several stages at which selection can occur, and th

bias favoring an initially proliferative response need not be as stringent. Evidently the

initial proliferation/differentiation ratio required to assure some minimum maturation po-

tential will depend upon the number of steps leading to plasma cell production.

7.2   The IgG Response

The general characteristics are similar to, but more pronounced than, those of th

IgM response. Specifically, maturation continues for about a month, during which time I

changes by nearly two orders of magnitude (Clafin, Merchant and Inman, 1973). This is

be compared to a nine-day maturation period and an order of magnitude change in I_{50} for

IgM (for the same antigen – DNP AGG). In both cases there is a trend toward dominance

high affinity clones whose precursors were present prior to immunization.

An important factor which must be considered in the comparison of IgG with IgM

hibition is the secretion rate distribution. I have speculated that early in the IgM respon

secretion rate is relatively restricted. This is not expected to be the case for IgG respon

in which differentiation has begun. Thus a substantial portion of the breadth of an inhi-

bition curve may reflect secretion rate. For example, with the secretion rate and affinity

distributions spanning two and four orders of magnitude, respectively, the observed inh

bition curve might span close to six (see Chapter 6). The breadths are approximately ac

ditive on a logarithmic scale and this raises some questions as to how the breadth of curv

obtained late in the IgG response should be interpreted. The data of Clafin et al. indicat

that at day 30, 80% of the plaques are inhibited within a two-decade interval. Consequen

if secretion rate spans two decades (see below), that does not leave much room for affinit

variation. Unfortunately, little is known about the secretion rate distribution, so further

speculation at this point is not warranted. It would clearly be profitable if experiments were performed under conditions outlined in the last chapter in order to isolate the factors which contribute to breadth of inhibition.

If the above remarks are correct, the changes in the breadth of the IgG and IgM inhibition curves cannot be unambiguously related to affinity maturation. In fact, the change in the difference in I_{50} between the two responses would actually be a minimum estimate of the difference in maturation potential, which may be quite substantial. The question is why a difference should exist at all.

One possibility is that the precursor cell population is more restricted for IgM than IgG. However, it is not clear why this would be so, since diversity involves V_H and V_L genes whereas class involves C_H genes. One would have to assume that there are some V_H and V_L genes in cells committed to IgM production that are never expressed (or even absent), and it is difficult to see how this would be advantageous. In fact, it would be simpler to assume that a cell initially expressing genes for V_L, V_H, C_L and C_μ, switches from C_μ to C_γ (C_μ and C_γ represent the constant regions of IgM and IgG, respectively); a strategy that is also consistent with a considerable amount of data, including the occurrence of cells secreting both IgG and IgM (Eisen, 1974). In addition, the inhibition results for both IgM and IgG indicate little, if any, difference in the breadth of curves obtained early in the respective responses.

An alternative explanation for the differences in maturation is to assume that the extent of selection is related to its duration. IgG matures more, simply because the response lasts longer. The question then becomes equivalent to asking why an IgG to IgM switch occurs when it does, the same question which arose in discussing the implications of intra-molecular reactions (Chapter 2). One can only speculate that some sort of balance must be maintained between the beneficial effects of a potent IgM response and the potential hazards of autoimmunity if the antigen cross reacts with self components.

There are a number of experiments in addition to those by Clafin et al. indicating good correlation between cellular maturation inferred from plaque inhibition, and serum maturation measured by Farr assays (e.g., Davie and Paul, 1972). However, North and Askonas (1974), in a study of monoclonal anti-DNP IgG responses by inhibition of plaque-forming cells, present evidence to the contrary. For their system, in any given inhibition

experiment, all cells are derived from the same clone and hence secrete identical antibo
The I_{50} values obtained from different clones of cells correlated poorly with the corresp
ing serum affinities measured by standard techniques.

The first question that should be considered in trying to understand these resu
is whether or not there are any fundamental differences between conditions of these exp
ments and those in which good correlation is obtained. Since the experiments study IgG
an important variable to consider is epitope density. In this case, RBC haptenation was
ried out with dinitrophenylated anti-sheep RBC antibody. Since there are generally sev
eral DNP groups per antibody, the epitope density can be considered high, even if the
antibodies themselves are sparsely distributed on the cell surface. Thus the theoretical
requirement for bivalent binding to RBC is probably satisfied. In addition, the inhibito
is DNP-lysine, so attachment to it is univalent, fulfilling a second theoretical requireme
Finally, North and Askonas present inhibition curves for both a heterogeneous and mon
clonal responses, showing about two orders of magnitude difference in breadth. Since
breadth of the latter reflects only secretion rate, the difference between the two must re
flect affinity heterogeneity. The experimental system therefore appears to be sensitive
affinity.

Since the preparative conditions appear to meet the conditions of the theory, equa
6.47 should be applicable. It predicts that on a plaque boundary, including that of the s
est observable plaque, $\frac{S}{K_1 H}$ is constant. Consequently, when all cells are from the sar
clone so that K_1 is constant, S/H must be constant, and this can only happen if S increa
as H. Therefore, with increasing H, higher secretion rate plaques will be inhibited an
breadth of the curve will reflect the secretion rate distribution.

Since $\frac{S}{K_1 H}$ is approximately constant on the smallest observable plaque radiu
it is evident that both secretion rate and affinity determine the midpoint of the inhibitio
curve. Thus the relationship between inhibitability and affinity only holds if the affini
difference between two populations is not compensated by a corresponding secretion ra
difference. Such secretion rate "interference" may explain the North-Askonas results,
if it does, the natural question is why this has not manifested itself in other experiment

The uniqueness of the North-Askonas experiments is in the comparison of mon clonal subpopulations selected from within a larger polyclonal population. This is to be contrasted with most other experiments which compare different polyclonal responses. If the difference in median secretion rates between the polyclonal responses is small compared to differences in median affinities, then a relationship between inhibitability and affinity should be observed (assuming, of course, that RBCs are properly prepared and an appropriate inhibitor is used). However, if within a polyclonal population, secretion rates span a wide range, secretion rate differences between two clones randomly picked from within the population may be sufficient to compensate for any affinity difference between them. This may be what is happening in the North-Askonas experiments. The extent to which such interference occurs will depend upon whether there is any correlation between affinity and secretion rate and, if so, how strong it is. In fact, the importance of the North-Askonas technique is that in conjunction with serum affinity measurements it allows a study of the relation between the dynamics of the affinity and secretion rate distribution.

Finally, the above considerations also suggest why good agreement has been found between the affinity of myeloma plaques and I_{50}. Since myelomas are clones of plasma cells, they not only secrete antibody with the same specificity, but they are all more or less differentiated to the same extent. Consequently, it would not be surprising if they all had similar secretion rates, in which case interference would arise.

7.3 Summary of Conclusions

I have tried to develop some of the biological inferences that can be drawn from the mathematical analysis of hemolytic plaque experiments. The approach has been intentionally speculative. My hope is that conjecture based upon reasonable supposition will stimulate further thought, and perhaps experimentation, in the directions indicated by the analysis. The principal conclusions follow.

(1) The preferred morphology of IgM free in solution is planar.

(2) The conformational freedom of the hinge in the IgMs is considerably more constrained than in IgG.

(3) The rate of IgM secretion from PFCs shows little variation up to at least four days following immunization. This may indicate that the initial phase of the response is

primarily proliferative.

(4) Most IgM maturation occurs during the first four days of the response.

(5) For IgG plaque-forming cells, the secretion rate distribution is broad, spa
ning at least two orders of magnitude.

(6) The maturation of the IgM response is probably considerably greater than
dicated by the change in I_{50}.

(7) The conditions on secretion rate that must be met to obtain correlation betw
inhibitability and affinity when comparing monoclonal populations may sometimes be dif
ent from those which must be met when polyclonal populations are being compared.

These conclusions are based upon the analysis of experiments in which respon
to DNP were studied. It is well known that qualitative as well as quantitative character
tics of a response can depend strongly on the nature of the antigen, including the exten
carrier haptenation. Therefore, the conclusions, if valid at all, are likely to be restric
to a class of antigens with some distinctive characteristics. Determining the immunolog
conditions (of which antigen geometry is just one) which lead to a response with certair
distinctive characteristics is a difficult problem that is currently under investigation.

8

DYNAMICAL PHENOMENA ON LYMPHOCYTE MEMBRANES

8.1   Background and Definitions

The concept of a fluid membrane (Singer and Nicholson, 1972) with laterally mobile lipid and protein constituents has had considerable impact upon recent directions in immunological experimentation. In the past six years much has been learned of the general organization of the lymphocyte plasma membrane, and there is currently considerable effort being expended to develop an understanding of the structure and function of its components. Of particular interest is the finding that cellular receptors can be redistributed upon binding antigen.

It is now known that incubating any of a variety of cell lines including lymphocytes, basophils and mast cells, with multifunctional ligands, may substantially perturb membrane organization. In particular, when B cells are stained with fluoresceinated anti immunoglobulins, the initially uniform, diffuse fluorescent pattern is observed after a few minutes to cluster into <u>patches</u> (Loor, Forni, Pernis, 1972; De Petris and Raff, 1973). These may continue to coalesce, eventually leading to complete localization over the Golgi apparatus (Taylor <u>et al</u>., 1971; Unanue <u>et al</u>., 1972; Yahahara and Edelman, 1972) where the fluorescent <u>cap</u> is subsequently endocytosed (<u>i.e</u>., ingested by the cell). Thus antigen binding to membrane receptors may be followed by three distinct dynamical processes: patching, capping and endocytosis. The first of these is apparently passive in the sense that it may occur in metabolically inhibited cells, whereas the latter two are energy dependent (Edidin and Weiss, 1972; Unanue, Karnooshy and Engers, 1973).

Patch formation is generally regarded as the two-dimensional analogue of three-dimensional aggregation such as precipitation, agglutination, and polymerization. The common characteristic of such processes is the formation of large branched networks. For example, a system initially composed of multifunctional monomers may begin to aggregate as the result of random collision and specific bond formation leading to dimer production. These may then react among themselves or with monomers forming tetramers and trimers, which in turn react to form larger aggregates. Thus, at any particular time there is a dis-

tribution of sizes which shifts in the direction of large complexes as time increases. Th
process is referred to as lattice formation.

The analogue of this on a membrane is receptor cross linking, i.e., bond form
between free sites on surface bound antigen and free receptor sites, mediated by diffus
in the plane of the membrane. However, the events on a metabolically active cell are m
complicated since, at some point in the aggregation process, ordered flow commences
sweeping the patches to a cell pole and facilitating further coalescence. Thus, during
early stages of the process, collision between sites involves only diffusion whereas at l
stages transport becomes important.

The relationship between these events and biological activity is not obvious. T
is evidence that cross linking on basophils is sufficient to trigger activation (histamine r
lease) and that, moreover, only a single cross linking step is necessary, i.e., only tw
receptors need be involved in the complex (Becker et al., 1973). There is also consid
able in vitro evidence that B cell activation (i.e., induction of plaque forming cells) by
T-independent antigens requires lattice formation (Greaves and Bauminger, 1972; Kish
moto and Ishizaka, 1975). Such antigens are usually large mitogenic polysaccharides v
a repeating array of determinants.

These observations imply that under certain circumstances, cross linking is a
necessary condition for activation. It need not, however, be sufficient. Since a cross
linked antigen is considerably more stable than one which is singly bound, cross link
may be required only to hold an antigen to a cell long enough for some nonspecific mite
signal to be delivered (Coutinho and Möller, 1974). Thus it would mediate delivery of
signal, but would not constitute the signal. If the mitogenic portion of an antigen were
missing, cross linking would occur without activation.

An alternative hypothesis might be that a certain minimum extent of cross link
is required to deliver a signal. For example, patch formation might open membrane ch
nels, allowing an influx of calcium ions, perhaps leading to changes in intercellular cy
AMP and GMP levels and thus altering metabolism.

Both these possibilities, signaling with and without patch formation, are plaus
and, moreover, they are not mutually exclusive. In fact, stimulation may lead to variou

combinations of differentiation and proliferation, depending upon conditions and the nature of the antigen, so it would not be surprising if there was more than one type of signal. In an attempt to disentangle some of the possibilities, a mathematical model for patching will be proposed and its predictions compared with known behavior. First, however, a digression into the theory of three-dimensional precipitation processes will be necessary.

8.2   The Dynamics of Antigen-Antibody Aggregation in Three Dimensions

8.2.1   The Aggregate Size Distribution Function

Consider a three-dimensional system consisting of bivalent antibodies at concentration $S_0/2$ and f-valent antigens at concentration L_0/f. If $\bar{S}$ and $\bar{L}$ are the concentrations of free antibody and antigen sites, respectively, then a parameter, p, called the extent of reaction can be defined as

$$p = \frac{S_0 - \bar{S}}{S_0} = \frac{L_0 - \bar{L}}{r\,L_0} \qquad\qquad 8.1$$

where

$$r \equiv \frac{S_0}{L_0} \qquad\qquad 8.2$$

The second equality in equation 8.1 arises because the number of bound antibody sites must equal the number of bound antigen sites. When all complexes are randomly distributed, p is also the probability that an antibody site picked at random is bound. One can also define a corresponding parameter for antigen sites. In particular, if p_1 is the probability that a randomly picked antigen site is bound, then from equation 8.1

$$p_1 = rp \qquad\qquad 8.3$$

With these definitions, an expression for the distribution of aggregate sizes as a function of p can be obtained.

In the absence of intramolecular reaction, the composition of an aggregate can be completely specified by the parameter pair (x,y), the number of antigens and antibodies in the aggregate. However, rather than use y, it will be more convenient to use the number of free antibody sites, n. This can always be done, for there is a simple constraint relating

n to y. The relation is obtained by recognizing that x – 1 antibodies must serve as connectors for x antigens and must therefore be bound at both sites. Since the total number
bound antibodies is the sum of those functioning as connectors plus those singly bound,

$$y = (x - 1) + n \qquad\qquad 8.4$$

Consequently, the composition of an aggregate is completely specified in terms of the pa
rameter pair (x,n).

We are interested in obtaining an expression for $C_n(x,p)$, the most probable cor
tration of aggregates with x antigens and n singly bound antibodies when the extent of rea
is p. A number of methods can be used to obtain this, provided one assumes that all fre
sites of the same type have equal a priori probability of reacting (see below). One met
is to write an expression for the total number of ways, Ω, of forming the aggregates C_n
for all allowed values of x and n, and then to maximize Ω with respect to $C_n(x,p)$ (Stoc
mayer, 1943). An alternative approach is to calulate directly P(x,n), the probability th
a free site picked at random is part of an (x,n)mer, for then

$$P(x,n) = \frac{\text{The number of free sites on (x,n)mers}}{\text{Total number of free sites}}$$

$$\qquad\qquad 8.5$$

$$= \frac{(fx - 2x + 2)\, C_n(x,p)}{\bar{L} + \bar{S}}$$

where $\bar{L}$ and $\bar{S}$ are the concentrations of free antigen and antibody sites, respectively (D
and Perelson, in press, 1976).

P(x,n) can be written as a sum of products; in particular,

$$P(x,n) = \sum_{i=1}^{2} \rho_i\, W_i\, \Omega_i \qquad\qquad 8.6$$

where i = 1 labels antigen, and i = 2, antibody. ρ_1 is the probability of picking a free an
gen site; W_1, the probability that the site is on an (x,n)mer, given that it is a free antig
site; and Ω_1 the number of ways of forming the (x,n)mer, again given that a free antige
site is picked. Analogous definitions hold for ρ_2, W_2 and Ω_2.

The expressions for ρ_1 and ρ_2 are easily written. By definition

$$\rho_1 = \frac{\bar{L}}{\bar{L} + \bar{S}} \qquad\qquad 8.7$$

and

$$\rho_2 = \frac{\bar{S}}{\bar{L} + \bar{S}} \qquad\qquad 8.8$$

Since $\bar{L}$ is the difference between the total antigen site concentration and the bound site concentration, then

$$\bar{L} = L_0 (1 - rp) \qquad\qquad 8.9$$

Similarly,

$$\bar{S} = S_0 (1 - p) \qquad\qquad 8.10$$

Hence

$$\rho_1 = \frac{L_0 (1 - rp)}{\bar{L} + \bar{S}} \qquad\qquad 8.11$$

and

$$\rho_2 = \frac{S_0 (1 - p)}{\bar{L} + \bar{S}} = \frac{rL_0 (1 - p)}{\bar{L} + \bar{S}} \qquad\qquad 8.12$$

The expression for W_1 can be written as a product of three terms:

(1) The probability that a specific sequence of x antigens is linked by $x - 1$ antibody connectors. This is $(p\, p_1)^{x-1}$.

(2) The probability that n antibodies are singly bound (i.e., have one free and one bound site). This is $[p_1 (1 - p)]^n$.

(3) The probability that all remaining antigen sites are free. Since there are $2(x - 1)$ antigen sites involved in holding the aggregate together, and another n are occupied by singly bound antibodies, that leaves $fx - 2(x - 1) - n$ free, one of which is the randomly picked site. Hence, the required term is

$(1 - rp)^{fx - 2x + 1 - n}$. Therefore

$$W_1 = (pp_1)^{x - 1} [p_1 (1 - p)]^n (1 - p_1)]^{fx - 2x + 1 - n} \qquad\qquad 8.13$$

Similarly,

$$W_2 = p\,(pp_1)^{x - 1} [p_1 (1 - p)]^{n-1} (1 - p_1)^{fx - 2x + 2 - n} \qquad\qquad 8.14$$

The combinatorial factor, Ω_1, can be written as the number of ways an $(x,0)$m can form, multiplied by the number of ways of distributing n singly bound antibodies an the remaining free sites in the $(x,0)$mer. However, the number of ways of forming an $(x$ mer is the same as the number of ways an x-mer can be formed in a system containing ϵ f-functional antigens that react among themselves when cycles are excluded. This foll by equating the $(x - 1)$ bonds in the latter system with the $(x - 1)$ antibodies in the for. The number of ways of forming an aggregate of x indistinguishable f-functional molecul which can interact among themselves is easily shown to be (Flory, 1953)

$$\frac{(f\,x - x)!}{(fx - 2x + 1)!\;x!}$$

To obtain Ω_1, this must be multiplied by the number of ways of distributing n bodies over the remaining $fx - 2x + 1$ free sites. This is

$$\frac{(fx - 2x + 1)!}{n!\;(fx - 2x + 1 - n)!}$$

Hence

$$\Omega_1 \;=\; \frac{(fx - x)!}{x!\;n!\;(fx - 2x + 1 - n)!} \qquad\qquad 8.15$$

Similarly,

$$\Omega_2 \;=\; \frac{(fx - x)!}{x!\;(n - 1)!\;(fx - 2x + 2 - n)!} \qquad\qquad 8.16$$

Substituting equations 8.11 – 8.16 into equation 8.6 and using equation 8.5, one has th (Goldberg, 1952)

$$C_n(x, p) = L_0 r^{x + n - 1}\, p^{2x - 2 + n}\,(1 - rp)^{fx - 2x + 2 - n}\,(1 - p)^n$$

$$x\,\frac{(fx - x)!}{(fx - 2x + 2 - n)!\;x!\;n!} \qquad\qquad 8.17$$

To complete the problem, p must be specified as a function of time. As an exa consider an irreversible, bimolecular reaction between antigen and antibody sites occur with forward rate constant k_1. Then

$$\frac{d\overline{S}}{dt} = -k_1 \overline{S} \overline{L} \qquad \overline{S} (t = 0) = S_0$$
$$\overline{L} (t = 0) = L_0$$

8.18

Since the concentrations of bound antigen and antibody sites must be equal,

$$S_0 - \overline{S} = L_0 - \overline{L}$$

8.19

and equation 8.18 becomes

$$\frac{d\overline{S}}{dt} = -k_1 \overline{S} (L_0 - S_0 + \overline{S})$$

8.20

the solution to which is

$$\overline{S} = \begin{cases} \dfrac{S_0 (1 - r) e^{-\lambda t}}{1 - r e^{-\lambda t}} & r \neq 1 \\[3em] \dfrac{S_0}{1 + S_0 k_1 t} & r = 1 \end{cases}$$

8.21

where

$$\lambda = L_0 (1 - r) k_1$$

Hence

$$p = \frac{S_0 - \overline{S}}{S_0} = \begin{cases} \dfrac{1 - e^{-\lambda t}}{1 - r e^{-\lambda t}} & r \neq 1 \\[3em] \dfrac{S_0 k_1 t}{1 + S_0 k_1 t} & r = 1 \end{cases}$$

8.22

and the distribution can be written as an explicit function of time.

It may be of interest to notice that what has in effect been done is to solve a system of coupled quadratic differential equations. For example, with $f = 2$, the differential equations which must be solved to obtain $C_0 (x,t)$, $C_1 (x,t)$ and $C_2 (x,t)$ are

$$\frac{d\,C_0(x,t)}{dt} = 2k_1 \sum_{j=1}^{x-1} C_1\,(x - j,\,t)\,C_0\,(j,t) - 2k_1\,\bar{S}\,C_0\,(x,t) \qquad\qquad x \geq 1 \qquad 8.23$$

$$\frac{dC_1(x,t)}{dt} = 4k_1 \sum_{j=1}^{x} C_2\,(x - j,\,t)\,C_0\,(j,t) + k_1 \sum_{j=1}^{x-1} C_1\,(x - j,t)\,C_1(j,t) \qquad 8.24$$

$$- k_1\,C_1\,(x,\,t)\,(\bar{S} + \bar{L}) \qquad\qquad x \geq 1$$

$$\frac{dC_2}{dt}\,(x,t) = 2k_1 \sum_{j=1}^{x} C_2\,(x - j,\,t)\,C_1\,(j,t) - 2k_1\,C_2\,(x,t)\,\bar{L} \qquad x \geq 0 \qquad 8.25$$

where $\bar{S}$ is given by equation 8.21 and $\bar{L}$ by 8.19.

For $f > 2$ and with reversible reactions, the system is much more complicated, graphical methods can be used to facilitate their derivation (Perelson and DeLisi, 1975) It is easily verified that the solution to equations 8.23 – 8.25 is given by equation 8.17 $f = 2$. In this particular case, it is not difficult to solve the kinetic equations directly by recursion. However, for arbitrary f, solving the system, or even writing it, is an imp ing problem and not a very fruitful way of finding the distribution.

Although it is not obvious from the above development, the theory has certain asp in common with those of non-ideal gas condensation and could, in fact, have been formula along similar lines (Stockmayer, 1943; Mayer and Mayer, 1940). One common feature o interest is the prediction of the existence of a critical parameter which signifies, in the pr pitation problem, an abrupt shift in the mass of the system from very small to very larg aggregates.

Consider the convergence of the sums

$$\sigma_j = \sum_{x=0}^{\infty} \sum_{n=0}^{fx - 2x + 2} x^{j}\,C_n\,(x,p) \qquad\qquad 8.26$$

These are related to the moments of the aggregate distribution function. The $j = 0$ term the total number of independent molecules in the system; $j = 1$ corresponds to the numbe average molecular weight (Tanford, 1961) of antigen in aggregates; $j = 2$ to the weight average molecular weight, etc. Summing over n, one has that

$$\sigma_j = \frac{L_0 (1 - rp^2)^2}{rp^2} \sum_{x=1}^{\infty} \frac{x^j Y^x (fx - x)!}{x! (fx - 2x + 2)!}$$

8.27

where

$$Y = rp^2 (1 - rp^2)^{f-2}$$

8.28

For large x

$$x! \sim \sqrt{2\pi x} \; e^{-x} x^x$$

Consequently, the coefficient of x^j in the sum goes asymptotically as

$$\left[\frac{Y (f-1)^{f-1}}{(f-2)^{f-2}} \right]^x x^{-5/2} \qquad f > 2$$

8.29

and

$$\frac{Y^x}{2} \qquad f = 2$$

Since a sum of the form

$$\sum_{x=1}^{\infty} x^{-k} Z^x$$

converges for $Z \leq 1$ if $k > 1$ and for $Z < 1$ if $0 \leq k \leq 1$, it is evident that the radius of convergence of σ_j is given by

$$Y_c = \frac{(f-2)^{f-2}}{(f-1)^{f-1}} \qquad f > 2$$

8.30

and at this point, σ_0 and σ_1 converge while σ_j, $j \geq 2$ diverges. When $f = 2$, the radius of convergence is

$$Y_c = 1$$

8.31

and σ_j converges only when $Y < 1$ for all j.

Defining $q = rp^2$

it is evident from equation 8.30 that the radius of convergence is reached at a value q = •
satisfying

$$q_c (1 - q_c)^{f-2} = \frac{(f-2)^{f-2}}{(f-1)^{f-1}} \qquad f > 2 \qquad \qquad 8.32$$

It can also be shown that for $0 \leq q \leq 1$ (which is the range of interest since $0 \leq rp \leq 1$ •
$0 \leq p \leq 1$), there is a unique real positive solution to this equation given by

$$q_c = \frac{1}{f - 1} \qquad \qquad 8.33$$

Because σ_2 just diverges at $q = q_c$, one can think of q_c as being related to the extent of
reaction at which the mass of the system shifts from predominantly small to predominant
large aggregates.

8.2.2   The Role of Diffusion

The equivalent site hypothesis assumes that all free sites have equal a priori p•
bability for reacting, regardless of the size of the aggregate to which they are attached
is evident, however, that as aggregates grow, they diffuse less readily, and one theref•
expects the time between encounters for the larger aggregates to be longer than for smal•
ones. In addition, because of chemical reaction, the free sites in the vicinity of an aggr•
gate will be depleted and hence the overall free site distribution will not be random. It tu.
out that neither of these effects is important so long as the chemical reaction is not diffusio•
limited (see below) and the viscosity is not unusually high.

With regard to the first effect, it is important to distinguish encounters from co•
ions. Sites are said to encounter one another when they come within a distance appropr•
for bond formation. At this distance there will generally be many collisions before a su•
cessful reaction. Although a decreased diffusion coefficient prolongs the time between
counters, it also prolongs the series of collisions at the encounter distance, since molec•
cannot diffuse away as readily. It can be shown formally that these two effects cancel •
a wide range of conditions (Flory, 1953).

As a simple model for evaluating effects of nonuniformity resulting from chemic•
reaction, consider two types of units A and B of radii, r_A and r_B, which may react whe•

their centers are separated by

$$r^* = r_A + r_B$$

Let a coordinate system be rigidly attached to A, with center at A's center of mass. Then C_B, the concentration of B at (r,t), satisfies

$$\frac{\partial C_B}{\partial t} = D \nabla^2 C_B$$

where D is the sum of the diffusion coefficients

$$D = D_A + D_B$$

The diffusion equation is to be solved subject to the boundary conditions that C_B be uniform far from A, i.e.,

$$\lim_{r \to \infty} C_B (r,t) = C_0$$

and that the flux across the reactive surface at $r = r^*$ be proportional to the reaction rate.

$$k_1 C_B (r^*, t) = \lim_{r \to r^*} 4\pi r^2 D \nabla C_B$$

The solution for these boundary conditions is, after a rapid transient,

$$C_B (r) = C_0 - \frac{k_1 C_B (r^*)}{4\pi Dr}$$

or, at $r = r^*$

$$C_B (r^*) = \frac{C_0}{1 + k_1/4\pi Dr^*}$$

Therefore the concentration at r^* is reduced by the factor

$$\frac{1}{1 + k_1/4\pi Dr^*}$$

and since the reaction is given by

$$k_1 C_B (r^*) = \frac{k_1 C_0}{1 + k_1/4\pi Dr^*}$$

it is evident that with the concentration taken as C_0, the effective forward rate constant will be

$$k_{eff} = \frac{k_1}{1 + k_1/4\pi Dr^*}$$

$4\pi Dr^*$ is the Smoluchowski (1917) diffusion-limited rate constant. So long as k_1 is not close to its diffusion-limited value, $k_{eff} = k_1$. Other aspects of the equivalent site hypothesis (i.e., the assumption that all free sites of the same type have equal a priori probability of reacting) are discussed by Flory (1953).

8.2.3   Aggregation as a Branching Process. Critical Behavior

Consider the following branching process where the solid circles represent anti-

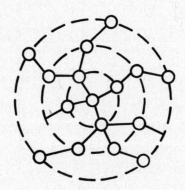

Figure 8.1

and the lines, bifunctional antibodies which connect them. The central circle represents an antigen picked at random. To keep things simple, cyclic reaction, including intramo-lecular bond formation, is prohibited, so the process can be represented topologically by a tree. The question of interest is: under what conditions will there be a nonzero proba-bility that the randomly picked antigen is part of an infinite aggregate?

Define a branching parameter, q', as the probability that a given antigenic site attached to a site on another antigen by an antibody connector. Then if Y_i is the number of antigens on the ith dashed circle, there are $(f - 1)Y_i$ sites which can branch to the $(i + 1)^{st}$. Hence, the expected number of antigens on the $(i + 1)^{st}$ circle is

$$Y_{i + 1} = q' (f - 1) Y_i \qquad\qquad 8.34$$

The actual number will of course fluctuate from this value, but for Y_i sufficiently large, the ratio of the actual number to the expected number should approach unity. Thus, since

$$\frac{Y_{i + 1}}{Y_i} = q' (f - 1) \qquad\qquad 8.35$$

infinite aggregate formation will not be possible if $\frac{Y_{i + 1}}{Y_i} < 1$. Consequently, the critical value of q' below which infinite aggregate formation is not possible is given by

$$q'_c = \frac{1}{f - 1} = q_c \qquad\qquad 8.36$$

To complete the identification of q' with q, notice that by definition q' is the probability that an antigenic site on circle i leads to an antigen on cirlce i + 1. It is therefore the product of the probability of a bound antigen site (rp) and a bound antibody site (p). Hence

$$q' = rp^2 = q \qquad\qquad 8.37$$

Equations 8.36 and 8.37 for critical branching are precisely the same as equation 8.33 obtained by studying the behavior of the σ_j. Evidently the critical value of the branching parameter represents the point at which large aggregate formation becomes possible. It has been interpreted as the point at which precipitation commences in antigen-antibody systems (Goldberg, 1952) and at which a sol to gel phase change occurs in polymerization processes (Flory, 1941). In addition, critical behavior of this sort is found in percolation processes (the diffusion of gas molecules through porous solids, Frisch and Hammersley, 1963; Shante and Kirkpatrick, 1971); epidemic theory (Hammersley, 1957) and non-ideal gas condensation (Mayer and Mayer, 1940). It can be shown that above the critical point the mass in all finite size aggregates is decreasing (Fischer and Essam, 1961), whereas below the critical point infinite aggregate formation is not possible.

Because of equation 8.33, large aggregate formation will be possible only if the composition of the system is within certain limits. Since

$$rp^2 > \frac{1}{f-1} \qquad \qquad 8.38$$

and with <u>antigen site excess</u>, the maximum value of p is one, then r_{min}, the minimum value of r for which equation 8.38 can be satisfied, is

$$r_{min} = \frac{1}{f-1}$$

so that

$$r > \frac{1}{f-1} \qquad \qquad 8.39$$

Alternatively, with antibody in excess, $(rp)_{max} = 1$ so that equation 8.38 implies

$$r < f-1 \qquad \qquad 8.40$$

Therefore (DeLisi, 1974)

$$\frac{1}{f-1} < r < f-1 \qquad \qquad 8.41$$

Outside this interval, unlimited growth is not possible.

Equation 8.33 is the condition which must be met to have a nonzero probability of unlimited growth, but it provides no information on what that probability is. An alternative derivation of the condition which provides more information is not difficult. Let Q be the probability that an antigen which is bound to at least one antibody is not part of an infinite aggregate; i.e., Q is the probability that an antigen does not coninue an infinite chain. Then

$$Q = (1-q)^{f-1} + (f-1)(1-q)^{f-2} qQ +$$

$$\frac{(f-1)}{2}(f-2)(1-q)^{f-3} q^2 Q^2 + \ldots + q^{(f-1)} Q^{f-1} \qquad \qquad 8.42$$

The first term on the right is the probability of no more bonds, the second the probability of one more bond times the probability that it does not continue an infinite chain, etc. Summing the right side, one has that

$$Q = [1 - q(1-Q)]^{f-1} \equiv g(Q) \qquad \qquad 8.43$$

The solutions to this equation will be at the intersection (s) of $x = Q$ with $x = g(Q)$.

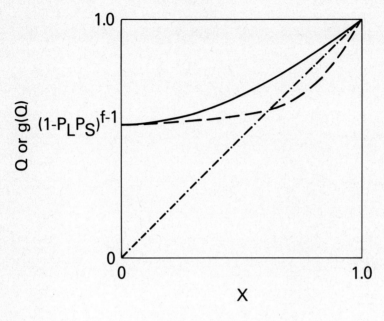

Figure 8.2

It is evident from Figure 8.2 that a necessary condition for a root, $x_0 < 1$, is that the slope of $g(Q)$ at $x = 1$ be greater than one.

$$\frac{\partial g}{\partial Q} = (f - 1) q (1 - q + qQ)^{f-2} \qquad 8.44$$

and

$$\frac{\partial g}{\partial Q}\bigg|_{x = 1} = (f - 1)q \qquad 8.45$$

Hence for

$$(f - 1) q > 1 \qquad 8.46$$

there is a root $x_0 \equiv Q_0 < 1$, whereas for

$$(f - 1) q \leq 1 \qquad 8.47$$

$Q \equiv 1$. Since $1 - Q$ is the probability that an antigen picked at random is part of an infinite aggregate, the first condition is the same as equation 8.36.

8.3 Critical Coalescence on Lymphocyte Membranes

Although the calculation of C_n (x,p) can be modified to treat the growth of aggregates on membranes (DeLisi and Perelson, 1976), there is no natural way to relate the distribution to a signal for triggering biological activity. Therefore, to be specific I will assume that the coalescence which occurs when $q - q_c$, fulfills a necessary condition for triggering. This also allows considerable simplification in the theory since the entire distribution need not be found in order to obtain an expression for the branching parameter.

When branching occurs on a surface, the argument relating q to valence is still valid. What changes is the relation between q and extent of reaction. The reason for the change is that in three-dimensional solutions some antigens may be entirely free, whereas for aggregation on a membrane all antigens must have at least one site bound. In addition, the membrane is not a closed system, since there may be a net influx of antigen onto the cell surface as aggregation proceeds. Define p_i as the probability that an antigen site picked at random is involved in a cross link. If L_0/f is the total concentration of membrane bound antigens, M' the concentration of multiply bound antigens, and M the concentration of bound sites on multiply bound antigens, then

p_i = probability that an antigen is multiply bound X probability that a randomly picked site on a multiply bound antigen is bound

$$= \frac{M'}{(L_0/f)} \times \frac{M}{fM'} = \frac{M}{L_0} \qquad 8.48$$

Since p_i plays a role analogous to p_1, the branching parameter will be given by

$$q = p_i p \qquad 8.49$$

where p_i is given by equation 8.48 and p by equation 8.1. The problem, therefore, is to obtain expressions for S, M, and L_0 as functions of time.

Consider the following reaction scheme. In the first step the antigen binds univalently and in the second it cross links (Bell, 1974). Intramolecular reaction, i.e., the

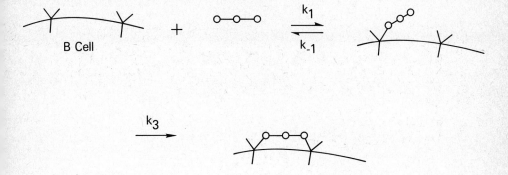

Figure 8.3. k_1 is the forward rate constant for interaction of an antigen molecule with a receptor site; k_{-1} the reverse rate constant for a site – site interaction; and k_3 the cross linking rate constant for a site – site interaction.

binding of antigen to two sites on the same molecule may also occur (DeLisi, 1976). However, the present model will neglect this possibility since it complicates the mathematics considerably. If the antigenic determinant density is sparse, intramolecular competition with cross linking should play a relatively minor role. However, if it is dense, omission of intramolecular bonds will lead to cross linking rates somewhat faster than they should be.

Even with this simplification, there are five variables to consider: the concentration of singly bound antigens, (m); the concentration of bound sites on multiply bound antigens, (M); the free receptor site concentration, $(\overline{S})$; the free antigen site concentration, $(\overline{L})$; and the total antigen site concentration, (L_0). Only three of these are independent, however, since there are two constraints. In particular, since the bound antigen site concentration must equal the bound antibody site concentration

$$L_0 - \overline{L} = S_0 - \overline{S} \qquad\qquad 8.50$$

In addition, by definition, the total concentration of bound antigen sites must be the sum of those on singly and multiply bound antigen.

$$L_0 - \overline{L} = m + M \qquad\qquad 8.51$$

With multiple binding considered irreversible within the time required for cellular signalling, the differential equations for m, $\bar{L}$ and L_0 are:

$$\frac{dm}{dt} = k_1 C \bar{S} - k_{-1} m - (f - 1) k_3 m \bar{S} \qquad 8.52$$

$$\frac{d\bar{L}}{dt} = -k_3 \bar{S} \bar{L} \qquad 8.53$$

$$\frac{dL_0}{dt} = f (k_1 C \bar{S} - k_{-1} m) \qquad 8.54$$

k_3 is the rate constant for cross linking and C is the concentration of free antigen in the vicinity of the cell surface. It is assumed to be in sufficient excess so that it does not ch appreciably during the course of the reaction.

In general, these equations must be solved numerically. However, if $\frac{dm}{dt}$ is cl to zero, one has that

$$k_1 C \bar{S} - k_{-1} m - (f - 1) k_3 m \bar{S} = 0 \qquad 8.55$$

Two limits may be considered, depending upon the relation between k_{-1} and $(f - 1) k_3$ If $k_{-1} \gg (f - 1) k_3 \bar{S}$, then

$$m \stackrel{\sim}{=} K_1 C \bar{S} \qquad 8.56$$

Substituting equation 8.56 into equation 8.54, one has that

$$\frac{dL_0}{dt} = 0 \qquad 8.57$$

or

$$L_0 = \text{constant} \qquad 8.58$$

According to equation 8.58, after a very rapid transient influx of antigen onto the cell s face, equilibrium is reached and L_0 assumes a constant value. The result is clearly a time approximation but it simplifies the problem considerably since it transforms the ce from an open to a closed system. Consequently, the constant can be evaluated by equili rium considerations. By definition

$$K_1 = \frac{S_0 - \bar{S}}{C\,\bar{S}} = \frac{L_0 - \bar{L}}{C\bar{S}} \qquad\qquad 8.59$$

Immediately after equilibrium with solution has been established, but before cross linking has begun, all antigens on the cell are singly bound; i.e., the concentration of bound antigen equals the concentration of bound antigenic sites.

$$\frac{L_0}{f} = L_0 - \bar{L} \qquad\qquad 8.60$$

Combining equations 8.59 and 8.60, one has that

$$L_0 = \frac{fKCS_0}{1 + KC} \qquad\qquad 8.61$$

and therefore

$$r \equiv \frac{S_0}{L_0} = \frac{1 + K\,C}{fKC} \qquad\qquad 8.62$$

Thus L_0 is determined and m and $\bar{S}$ are easily related. Moreover, $\bar{L}$ (and hence $\bar{S}$) can now be determined from equation 8.53. Substituting equation 8.50 into 8.53

$$\frac{d\bar{L}}{dt} = -k_3\,\bar{L}(S_0 + \bar{L} - L_0) \qquad\qquad 8.63$$

This is to be solved subject to the initial condition that at $t = 0^+$; i.e., as soon as equilibrium with solution antigen is established, all antigens are singly bound and hence the number of free antigenic sites is

$$\bar{L}\,(t = 0^+) = L_0 - \frac{L_0}{f} \qquad\qquad 8.64$$

With this condition, equation 8.63 integrates to

$$\bar{L} = \begin{cases} \dfrac{L_0 \dfrac{(f-1)}{f}\,(r-1)\,e^{-\mu t}}{r - 1 + \dfrac{(f-1)}{f}(1 - e^{-\mu t})} & r \neq 1 \\[4ex] \dfrac{L_0\,(f-1)}{f + L_0\,k_3\,(f-1)\,t} & r \neq 1 \end{cases} \qquad\qquad 8.65$$

where

$$\mu \equiv k_3 L_0 (r - 1) \qquad\qquad 8.66$$

An equation for p_i as a function of time can be obtained by using equations 8.50, 8.51 and 8.56 to express M in equation 8.48 as a function of $\bar{L}$. Then

$$p_i = r K_1 C [f (1 - \frac{\bar{L}}{L_0}) - 1] \qquad\qquad 8.67$$

and the branching parameter is

$$q = K_1 C (1 - \frac{\bar{L}}{L_0}) [f (1 - \frac{\bar{L}}{L_0}) - 1] \qquad\qquad 8.68$$

Since large aggregate formation becomes possible when equation 8.36 is satisfied, a criti[cal] time $(t = t_c)$ can be found at which q given by equation 8.68 equals $1/(f - 1)$. In particul[ar]

$$t_c = \frac{1 + K_1 C}{S_0 k_3 [1 - (f - 1) K_1 C]} \ln \left\{ \frac{1 - u_c}{\frac{(f - 1)}{f} [1 - (fu_c - 1) K_1 C]} \right\}, \quad r \neq 1 \qquad 8.69$$

where

$$u_c = \frac{1}{2f} + \frac{1}{2} \left[\left(\frac{1}{f}\right)^2 + \frac{4}{f(f - 1)K_1 C} \right]^{\frac{1}{2}} \qquad\qquad 8.70$$

The development of these equations is based upon the assumption that within tim[e] of interest, k_{-1} is large compared to $(f - 1) k_3 \bar{S}$; i.e., the dissociation rate is much fas[ter] than the cross linking rate. It is only because of this relationship that the binding react[ion] is sensitive to the equilibrium constant. In the opposite limit, cross linking occurs rapi[dly] on the time scale of dissociation. In this case,

$$m = \frac{k_1 C}{(f - 1) k_3} \qquad\qquad 8.71$$

However, closed form solutions still cannot be obtained. Nonetheless, it is evident that binding is no longer affinity-dependent, since the reverse reaction does not have a chan[ce] to occur.

8.4 Immunological Implications of the Model

Cellular activation is undoubtedly a very complex phenomen depending as it does not only upon antigen geometry, dose and chemical properties, but also upon a variety of host cells and molecules. It is not necessary, however, that all these factors need be considered explicitly in interpreting each aspect of an immune response. For example, if one considers the antigen-cell interaction, a variety of events may ensue, including: (1) dissociation, (2) intramolecular reaction, (3) cross linking with large aggregate formation, and (4) cross linking without large aggregate formation. These are fundamental physical chemical events and one can argue that the various ancillary host cells and molecules modulate, block or amplify one or more of them in various ways by affecting the rate constants. Thus the biochemical complexities are, in effect, parameterized.

Of course, one still does not know which combination of physical chemical events may be prerequisite to a biological phenomenon. Therefore, some insight may be gained by assuming a particular physical chemical correlate of some aspect of activation, pursuing the consequences, and comparing the predictions with observation. Since there is some indirect evidence that cross linking is required for B cell triggering, and no evidence that triggering can occur in the absence of cross linking, it is not unreasonable to start by assuming that triggering requires some minimum amount of cross linking to occur in a given amount of time. To be specific, a positive correlation between t_c (equation 8.69) and activation will be assumed, where activation is to be interpreted in the broadest sense, meaning both proliferation and differentiation.

When the dimensionless time $S_0 k_3 t_c$ is plotted against $\ln(K_1 C)$ for various values of f, one obtains

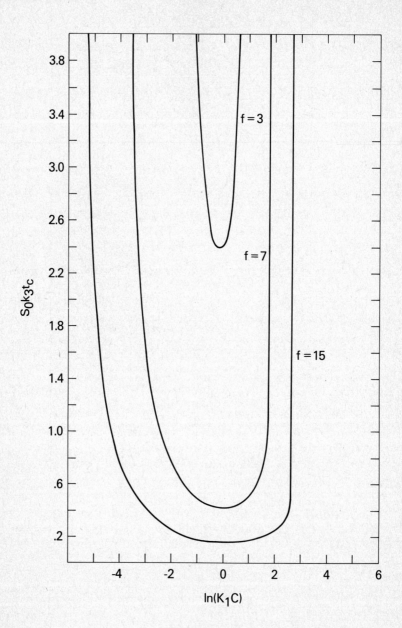

Figure 8.4

These results have the following features:

(1) As valence decreases, the critical time for coalescence increases. Moreover, the minimum time for coalescence also decreases with increasing valence. This suggests that there may be a minimum valence below which triggering will not be possible. For example, if triggering required $S_0 k_3 t_c \geq 1.0$, the antigens with $f = 7$ and $f = 3$ will not trigger unless they are aided by helper cells and molecules which effectively increase the valence.

(2) There is an optimum value of $K_1 C$ for triggering which occurs at about $K_1 C = 1$. This means that as C decreases, the optimum K_1 increases. Hence immune maturation is predicted.

As I already indicated, this prediction arises from the assumption that dissociation occurs rapidly compared to cross linking. Thus, for antigens which react with receptors according to this scheme, there will be maturation. Conversely, if there is maturation, the interaction should be representable, at some level, by a scheme of this sort. In the alternative limit (equation 8.71) it appears on physical grounds that there will be no maturation. However, mathematical demonstration of this and a detailed general study of the model could be profitable. Perhaps those T-independent mitogens which do not elicit a response that matures conform to this kinetic pattern.

(3) Each curve has two vertical asymptotes, Since

$$\lim_{t \to \infty} \frac{L}{L_0} = \begin{cases} 1 - r & r < 1 \\ 0 & r \geq 1 \end{cases}$$

the branching parameter has a definite maximum. Using the limiting value of $\frac{L}{L_0}$ and equating q to $\frac{1}{f-1}$ at this value, one finds that

$$K_1 C < f - 1 \qquad r < 1$$

and

$$K_1 C > \frac{1}{(f-1)^2} \qquad r \geq 1$$

Hence for values of $K_1 C$ which are either too low or too high, triggering will not be possible.

The minimal model thus distinguishes binding reactions from triggering reaction and predicts affinity maturation, unresponsiveness to high antigenic doses, and auxiliar cell requirements for low valence antigens.

8.5   Future Problems

There are a number of unsolved mathematical problems which have arisen, and would like to recall just a few which I think can be profitably investigated.

(1) A problem of fundamental scientific importance, which arose in Chapter 2, is to find the distribution function governing the distance between two units in a chain molecule (Section 2.2). The problem is analogous to a three-dimensional random walk w constant step size and has been solved exactly only when the links of the chain are connected by universal joints, so that for any step all directions are equally probable. The next simplest model is to allow a bond to rotate freely about the direction of its predecessc but to require a fixed angle between bonds (Figure 2.3). The question is: what is the pro bability that two units separated by N links will be a distance r apart? Progress toward t solution of problems of this sort is of interest in connection with a host of biochemical situ- ations in which one must assess the relation between cyclization reactions (i.e., the proba bility of return to the origin) and bimolecular reactions.

(2) The problem in Chapter 3, viz., that of finding the free energy distribution from binding data, still has not been adequately treated. The problem is to solve the int gral equation for the least restrictive analytical function [f(H)] which satisfies the physi requirements outlined in Section 3.2. Problems involving data convolution similar to thi arise in numerous other areas of biology (e.g., finding the distribution of lengths of cha molecules eluted from a chromatography column). Investigating these areas and formulati and solving the problem in as general a way as possible would therefore be of considerab importance.

(3) There are numerous problems involving the plaque assay, not the least of wh is to find whether a sound mathematical basis can be provided for the physical approximati introduced in Chapter 5. There are, in addition, questions related to the effects produc by single source secretion rate variations, and complement kinetics which was not treate

explicitly. The last two problems have strong experimental components and are undoubtedly best approached in collaboration with experimentalists. Diffusion reaction phenomena, of course, play a prominent role in biological reactions, but the plaque assay is one of the few in which the linear regime is of prime interest (with the possible exception of subthreshold phenomena in nerve conduction) and therefore represents an excellent system for detailed mathematical investigation and experimental collaboration. A study of the status of immuno-diffusion is given in Table 4.1.

(4) The area of B cell activation represents a fertile field for modeling, both with regard to dynamical processes on membranes (for example, Blumental, 1975) and cell-cell interaction (for example, Hoffman, 1975). Equations 8.52 - 8.54 represent a specific model illustrating how membrane phenomena can play a role in determining cellular behavior. However, although the results of Chapter 8 suggest some interesting biological possibilities, the behavior of these equations still needs thorough investigation.

BIBLIOGRAPHY

Abromowitz, M., and Segun, I. A. (1965) Handbook of Mathematical Functions. New Yor
    Dover Publications

Aladjem, F. and Palmiter, M.T. (1965) J. Theor. Biol. 33, 339

Aladjem, F., Klostergaard, H. and Taylor, R.W. (1962) J. Theor. Biol. 3, 134

Andersson, B. (1970) J. Exp. Med. 132, 77

Becker, K.E., Ishizaka, T., Metzger, H., Ishizaka, K. and Grimley, P.M. (1973) J. Ex
    Med., 138, 394

Bell, G.I. (1974) Nature 248, 430

Bell, G.I. (1975a) in Proc. First Los Alamos Life Sciences Symposium. USERDA Technic
    Information Center, Oak Ridge

Bell, G.I. (1975b) Transplant. Rev. 23, 23

Blumental, R. (1975) J. Theor. Biology 49, 219

Briggs, H.E. and Haldane, J.B.S. (1925) Biochem. J. 19, 338

Burnet, F.M. (1967) Cold Spring Harbor Symposium on Quant. Biol. 32, 1

Burnet, F.M. (1957) Austr. J. Sci. 20, 67

Cann, J.R. (1975) Immunochem. 12, 473

Carslaw, H.S. and Jaegar, J.C. (1959) Conduction of Heat in Solids. London: Oxford
    University Press

Clafin, L., Merchant, B. and Inman, J. (1973) J. Immunol. 110, 241

Clafin, L., Merchant, B. and Inman, J. (1972) Cellular Immunol. 5, 209

Coutinho, A. and Möller, G. (1974) Scand. J. Immun. 3, 133

Crothers, D.M. and Metzger, H. (1972) Immunochem. 9, 341

Cunningham, A. (1965) Nature 207, 1106

Davie, J. and Paul, W.E. (1972) J. Exp. Med. 135, 660

Day, L.A., Sturtevandt, J.M. and Singer, S.J. (1963) Ann. N.Y. Acad. Sci. 103, 611

DeLisi, C. (1974) J. Theor. Biol. 45, 555

DeLisi, C. (1975a) J. Theor. Biol. 51, 336

DeLisi, C. (1975b) J. Theor. Biol. 52, 419

DeLisi, C. (1975c) J. Math. Biol. 2., 317

DeLisi, C. (1976) "Antigen Binding to B Cell Immunoglobulin Receptors" in Theoretical
    Immunology, G.I. Bell, A. Perelson and G. Pimbley (eds.), in press.

DeLisi, C. and Bell, G.I. (1974) Proc. Nat. Acad. Sci. U.S.A. 71, 16

DeLisi, C. and Crothers, D.M. (1971) Biopolymers 10, 1809

DeLisi, C. and Crothers, D.M. (1973) Biopolymers, 12, 1689

DeLisi, C. and Goldstein, B. (1975) J. Theor. Biol. 51, 313

DeLisi, C. and Perelson, A. (1976) J. Theor. Biol., in press

De Petris, S. and Raff, M.C. (1973) Nature, New Biology 241, 257

Dutton, R.W. (1975) Transplant. Rev. 23, 66

Edidin, M. and Weiss, A. (1972) Proc. Nat. Acad. Sci. U.S.A. 69, 2456

Edmundson, A.B., Ely, K.R., Abola, E.E., Schiffer, M., Westholm, F.A., Fausch, M.D. and Deutch, H.F. (1974) Biochem. 13, 3816

Eisen, H. (1974) Immunology. Hagerstown, Md.: Harper and Row

Eisen, H.N. (1964) Methods in Medical Research. 10, 106

Eisen, H.N. and Siskind, G.W. (1964) Biochem. 3, 996

Erdélyi, A. (editor) (1954) Tables of Integral Transforms. New York: McGraw-Hill, p. 213

Feinstein, A., Munn, E.A. and Richardson, N.E. (1971) Ann. N.Y. Acad. Sci. 190, 104

Fischer, M.E. and Essam, J.W. (1961) J. Math. Phys. 2, 609

Flory, P.J. (1953) Principles of Polymer Chemistry. Ithaca, N.Y.: Cornell University Press

Flory, P.J. (1941) J. Amer. Chem. Soc. 63, 3083

Flory, P. (1969) The Statistical Mechanics of Chain Molecules. New York: Interscience, Ch. IV

Frisch, H.L. and Hammersley, J.M. (1963) SIAM. J. Appl. Math. 11, 894

Froese, A., Sehon, A.H. and Eigen M. (1962) Canad. J. Chem. 40, 1786

Goidel, E.A., Paul, W.E., Siskind, G.W. and Banaceraf, B. (1968) J. Immunol. 100, 371

Goldberg, R. (1952) J. Amer. Chem. Soc. 74, 5715

Goldstein, B. and Perelson, A. (1976) submitted

Goldstein, B., DeLisi, C. and Abate, J. (1975) J. Theor. Biol. 52, 317

Goldstein, B. and DeLisi, C. (1975) Immunochem. 13, 42

Graber, P. and Williams, C.A. (1953) Biochim. and Biophys. Acta 10, 193

Grad, H. (1949) Communic. Pure and Appl. Math. 2, 331

Greaves, M.F. and Bauminger, S. (1972) Nature, New Biology 235, 67

Hammersley, J.M. (1957) Proc. Cambridge Phil. Soc. 53, 642

Harel, S., Ben Efraim, S. and Liacopoulas, P. (1970) Immunol. 19, 319

Heineken, F.G., Tsuchiya, H.M. and Aris, R. (1967) Math. Biosciences 1, 95

Hill, T.L. (1960) An Introduction to Statistical Thermodynamics. Reading, Mass.: Addi
Wesley

Hiramoto, R.N., McGhee, J.R., Hurst, D.C., Hamlin, N.M. (1970) Immunochem. 7, 97:
8, 355

Hoffman, G. (1975) Eur. J. Immunol. 5, 638

Hornick, C. and Karush, F. (1969) Israel J. Med. Sci. 5, 163

Hornick, C. and Karush, F. (1972) Immunochem. 9, 325

Hughes-Jones, N.C., Gardner, B. and Telford, R. (1964) Immunology 7, 1972

Humphrey, J.H. (1967) Nature 216, 1295

Ingraham, J.S. (1963) C. R. Acad. Sci. 256, 5005

Ingraham, J.S. and Bussard, A. (1964) J. Exp. Med. 119, 667

Inman, J.K. (1974) in The Immune System: Genes, Receptors, Signals, E.E. Secua, A.
Williamson and C.F. Fox, eds. New York: Academic Press

Inman, J.K., Merchant, B., Clafin, L. and Tacey, S.E. (1973) Immunochem. 10, 165

Jerne, N.K. (1955) Proc. Nat. Acad. Sci. U.S.A. 41, 849

Jerne, N.K., Nordin, A.A. and Henry, C. (1963) The Agar Plaque Technique for Reco}
nizing Antibody Producing Cells, in Cell Bound Antibodies. Philadelphia: Wis
Institute Press, p 109

Jerne, N.K., Henry, C., Nordin, A.A., Fuji, H., Koros, A.M.C. and Lefkovits, I. (197
Transplant. Rev. 18, 131

Jernigan, R. (1967) Ph.D. Thesis. Stanford University

Kabat, E.A. (1968) Structural Concepts in Immunology and Immunochemistry. New Yor
Holt, Rinehart and Winston

Kishimoto, T. and Ishizaka, K. (1975) J. Immun. 114, 585

Laurell, C.B. (1966) Annal. Biochem. 15, 45

Loor, F., Forni, L. and Pernis, B. (1972) Eur. J. Immunol. 2, 203

Mancini, G., Carbonara, A.O. and Heremans, J.R. (1965) Immunochem. 2, 235

Mayer, J.E. and Mayer, M.G. (1940) Statistical Mechanics. New York: John Wiley and
Sons

Metzger, H. (1974) Advances in Immunology 18, 169

Michaelis, L. and Menton, M. (1913) Biochem. Z. 49, 333

Möller, Göran (ed.) (1975) Transplant. Rev. vol. 23

North, J.R. and Askonas, B.A. (1974) Eur. J. Immunol. 4, 361

Pasanen, V.J. and Mäkelä, O. (1969) Immun. 16, 399

Pecht, I., Givol, D. and Sela, M. (1972) J. Mol. Biol. 68, 241

Perelson, A. and DeLisi, C. (1975) J. Chem. Phys. 62, 4053

Raleigh, L. (1919) Phil. Mag. [6], 37, 321

Segal, D.M., Padlan, E.A., Cohen, G.H., Rudikoff, R., Pottor, M. and Davies, D.R.
    (1974) Proc. Nat. Acad. Sci. U.S.A. 41, 1427

Shante, V.K.S. and Kirkpatrick, S. (1971) Adv. Phys. 20, 325

Singer, S.J. and Nicholson, G.L. (1972) Science 175, 720

Sips, R. (1948) J. Chem. Phys. 16, 490

Siskind, G.W. and Banaceraf, B. (1969) Adv. in Immun. 10, 1

Smoluchowski, M.V. (1917) Zeitschr. Physik. Chemie. 92, 129

Steiner, L.A. and Eisen, H.N. (1967) J. Exp. Med. 126, 1164

Stockmayer, W. (1943) J. Chem. Phys. 11, 45

Suzuki, T. and Deutch, H.F. (1967) J. Biol. Chem. 242, 2725

Tanford, C. (1961) Physical Chemistry of Macromolecules. New York: John Wiley and Sons

Taylor, R.B., Duffus, W.P.H., Raff, M.C. and de Petris, S. (1971) Nature, New Biology
    233, 225

Unanue, E.R., Perkins, W.D. and Karnovsky, M.J. (1972) J. Exp. Med. 136, 885

Unanue, E.R., Karnovsky, M.J. and Engers, D.H. (1973) J. Exp. Med. 137, 675

Varga, J.M., Koningsberg, W.H. and Richards, F.F. (1974) Proc. Nat. Acad. Sci. U.S.A.
    70, 3269

Vitetta, E. and Uhr, J.W. (1975) Science 189, 964

Weiler, E., Melletz, E.W. and Breuninger-Peck, E. (1965) Proc. Nat. Acad. Sci. U.S.A.
    54, 1310

Werblin, T.P. and Siskind, G.W. (1972a) Transplant. Rev. 8, 104

Werblin, T.P. and Siskind, G.W. (1972b) Immunochem. 9, 987

Williams, C.A. and Chase, M.W. (eds.) (1971) Methods in Immunology and Immunochem.
    New York: Academic Press, vol. 3, p. 103

Wu, C.Y. and Cinader, B. (1971) J. Immunol. Methods 1, 19

Wu, T.T. and Kabat, E.A. (1970) J. Exp. Med. 132, 211

Yahara, I. and Edelman, G.M. Proc. Nat. Acad. Sci. U.S.A. (1972) 69, 608

Yguarabide, J., Epstein, H.F. and Stryer, L. (1970) J. Mol. Biol. 51, 573

INDEX

Biomathematics

Editors: K. Krickeberg;
S. Levin; R. C. Lewontin;
J. Neyman; M. Schreiber

Springer-Verlag
Berlin
Heidelberg
New York

Journal of

Mathematical Biology

Edited by
H.J. Bremermann,
Berkeley, Calif.
F.A. Dodge, Jr.,
Yorktown Heights, N.Y.
K.P. Hadeler
Tübingen

After a period of spectacular progress in pure mathematics, many mathematicians are now eager to apply their tools and skills to biological questions. Neurobiology, morphogenesis, chemical biodynamics and ecology present profound challenges. The **Journal of Mathematical Biology** is designed to initiate and promote the cooperation between mathematicians and biologists. Complex coupled systems at all levels of quantitative biology, from the interaction of molecules in biochemistry to the interaction of species in ecology, have certain structural similarities. Therefore theoretical advances in one field may be transferable to another and an interdisciplinary journal is justified.

Subscription information and sample copy available upon request.

FRG, West Berlin and GDR:
Please send your order or request to
Springer-Verlag, D-1000 Berlin 33, Heidelberger Platz 3

Rest of the World (excluding North America):
Please send your order or request to
Springer-Verlag, A-1011 Wien, Mölkerbastei 5

Springer-Verlag Wien New York

Springer-Verlag Berlin Heidelberg New York

This series aims to report new developments in biomathematics research and teaching – quickly, informally and at a high level. The type
of material considered for publication includes:

1. Preliminary drafts of original papers and monographs

2. Lectures on a new field, or presenting a new angle on a classical field

3. Seminar work-outs

4. Reports of meetings, provided they are

   a) of exceptional interest and

   b) devoted to a single topic.

Texts which are out of print but still in demand may also be considered
if they fall within these categories.

The timeliness of a manuscript is more important than its form, which
may be unfinished or tentative. Thus, in some instances, proofs may be
merely outlined and results presented which have been or will later be
published elsewhere. If possible, a subject index should be included.
Publication of Lecture Notes is intended as a service to the international
scientific community, in that a commercial publisher, Springer-Verlag,
can offer a wider distribution to documents which would otherwise
have a restricted readership. Once published and copyrighted, they can
be documented in the scientific literature.

Manuscripts

Manuscripts should comprise not less than 100 pages.
They are reproduced by a photographic process and therefore must be typed with extreme care. Symbols
not on the typewriter should be inserted by hand in indelible black ink. Corrections to the typescript
should be made by pasting the amended text over the old one, or by obliterating errors with white cor-
recting fluid. Authors receive 75 free copies and are free to use the material in other publications. The
typescript is reduced slightly in size during reproduction; best results will not be obtained unless the
text on any one page is kept within the overall limit of 18 x 26.5 cm (7 x 10½ inches). The publishers
will be pleased to supply on request special stationery with the typing area outlined.

Manuscripts in English, German or French should be sent to Dr. Simon Levin, Center for Applied
Mathematics, Olin Hall, Cornell University Ithaca, NY 14850/USA or directly to Springer-Verlag Heidelberg

Springer-Verlag, Heidelberger Platz 3, D-1000 Berlin 33
Springer-Verlag, Neuenheimer Landstraße 28–30, D-6900 Heidelberg 1
Springer-Verlag, 175 Fifth Avenue, New York, NY 10010/USA

ISBN 3-540-07697-2
ISBN 0-387-07697-2